BEGINNING SCIENCE

PHYSICS

J J Wellington

Oxford University Press

Oxford University Press, Great Clarendon Street, Oxford OX2 6DP

Oxford New York
Athens Auckland Bangkok Bogota Bombay
Buenas Aires Calcutta Cape Town Dar es Salaam
Delhi Florence Hong Kong Istanbul Karachi
Kuala Lumpur Madras Madrid Melbourne
Mexico City Nairobi Paris Singapore
Taipei Tokyo Toronto

and associated companies in
Berlin Ibadan

Oxford is the trade mark of Oxford University Press

© J. J. Wellington, 1984

First published 1984
Reprinted 1985, 1986, 1988, 1989 (twice), 1991, 1993, 1994, 1996, 1997

ISBN 0 19 914093 6

Cover photographs:

Front: ZEFA Picture Library
Back: Paul Brierley (top)
      IBM United Kingdom Limited (bottom)

Typeset by Rowland Phototypesetting Ltd
Bury St Edmunds, Suffolk
Printed at The Bath Press, Bath

Contents

Galileo Galilei

Nicolaus Copernicus

John Dalton

Sir Isaac Newton

Robert Brown

Albert Einstein

Ernest Rutherford

Introduction

What is physics?

Physics is all about the things and happenings that you notice around you: lightning, fog, the tides, seeing and hearing, and many other things. Discoveries in physics have led to the invention of thousands of machines that affect your everyday life. Electricity, television, transport, robots, and electronics are a few examples.

Everything is made of some sort of material, or *matter*. Matter can be solid, liquid, or gas.

How and why

Physics tries to explain *how* things work or *why* things happen. 'How' questions are usually easier to answer. The laws of physics can describe how objects fall, how light travels, how a rainbow is made, or how a telescope works. 'Why' questions are often much harder: why does an apple fall? why does light travel in straight lines? why do magnets attract? why does water boil?

This book has been divided into six main Topics: Matter, Energy, Forces, Waves, Atoms and Electrons, and Fields. Each topic explains an important idea used in physics to explain how and why things happen.

Laws and theories   People who study physics are called physicists. Physicists spend a lot of time doing experiments which involve accurate measurements and recording of results. Sometimes the results of experiments follow a pattern or rule. If a pattern can be found it is written down as a *law*.

Some of the laws of physics were discovered hundreds of years ago and are still used today. Examples are Archimedes' principle (250 BC) and Newton's laws (1686). But laws in physics sometimes change as new discoveries are made.

The results of studies in physics are often applied to everyday life. One example is changing the *energy* of hot geysers into electrical energy, or electricity.

The laws of physics tell us *how* things happen. *Theories* are used to explain *why* they behave like they do.

Special words   In physics, many words have a special meaning. For example, the word 'work' has many different meanings in everyday life, from driving a bus to reading a physics book. In physics, 'work' has only one *exact* meaning. This is why physics is called an *exact science*. It deals with things or quantities that can be accurately defined and measured, such as work, speed, forces, heat, and electric current. Calculations are used in physics to work out or predict one quantity from others. These calculations can only be done with quantities that can first be measured exactly.

The history of physics

Ancient Greeks   The first physicists were probably the ancient Greeks. The most famous, Aristotle, was born in 384 BC. He suggested laws about how bodies move, about the Earth and the planets, and about the 'elements' that everything is made up of. Another ancient Greek, Archimedes, first explained why objects float or sink.

Physicists study *forces*. Tremendous forces are needed to launch a communications satellite into orbit.

The 17th century Many of Aristotle's ideas were believed right up to the time of Galileo (1564–1642). Galileo was the first physicist to carry out careful experiments. He studied the way that objects fall and made new laws of motion.

The most famous physicist of all, Sir Isaac Newton, was born in 1642. His discoveries about light, gravity, moving objects, and forces are still used in physics today.

Electricity The next landmark in physics came when electricity was 'discovered'. Once Volta had invented the electric cell or battery in 1800 scientists were able to do experiments with electricity and find new laws. Ohm, Ampère, and Oersted were three famous scientists who worked with electricity. But the most important discoveries were made by Michael Faraday, who first found the link between electricity and magnetism in 1831. His work led to the motors, electromagnets, and dynamos which people use everywhere today.

New physics It was Faraday who first thought of the idea of a 'field of force' around a magnet. The idea of fields is now used in the 20th century to help explain magnetism and gravity. Albert Einstein is probably the most famous physicist of this century. He tried all his life to explain the pulls of both gravity and magnetism with one single theory, but failed.

Atomic physics Most of the physics in the 20th century has been concerned with the tiny 'building bricks' that everything is made up of – *atoms*. The physics of the atom began in earnest with Ernest Rutherford when he 'split' the atom in 1911. Atomic and nuclear physics is now one of the most important areas of study. It has led to the nuclear reactor, invented by Fermi in 1942, and to the atomic bomb, first used in 1945.

Some famous names in physics and a summary of their discoveries are contained in the table on the next page.

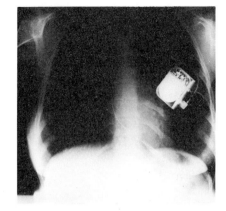

X-rays are part of the family of *waves* called the electromagnetic spectrum. This X-ray of a human chest shows a pacemaker implanted in the heart.

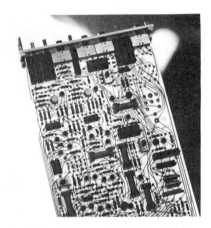

The study of *atoms and electrons* led to the discovery of electronics and computers. This photo shows a computer program panel.

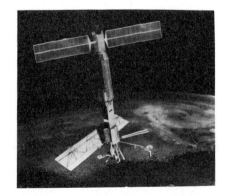

One type of *field* is a gravitational field, or gravity. The force from the Earth's gravity keeps a satellite in orbit.

Summary

1   Physics is all about the things that happen around you. Physics affects your everyday life in thousands of ways.

2   Important ideas in physics are matter, energy, forces, waves, atoms, and fields.

3   The laws of physics describe how things happen – they describe a rule or pattern. Theories are used to explain why they happen.

4   The history of physics extends from Aristotle to Galileo, Newton, and Faraday, up to the famous names of this century: Rutherford, Einstein, and Fermi.

Some famous names in physics

name	lived	nationality	famous for . . .
Democritus	470–400 BC	Greek	suggesting the 'atom'
Aristotle	384–322 BC	Greek	the first laws of motion
Archimedes	287–212 BC	Greek	laws of floating and sinking; the screw
Ptolemy	2nd century AD	Greek	believing Earth to be the centre of the Universe

★★★★

name	lived	nationality	famous for . . .
Nicolaus Copernicus	1473–1543	Polish	picturing the universe with the Sun at the centre
William Gilbert	1544–1603	English	studying magnets
Galileo	1564–1642	Italian	first experiments, telescope, and the laws of falling bodies
Johannes Kepler	1571–1630	German	laws of planetary motion
Robert Hooke	1635–1703	English	Hooke's law for springs
Sir Isaac Newton	1642–1726	English	the spectrum, and three new laws of motion
Thomas Newcomen	1663–1729	English	inventing the steam engine
Daniel Fahrenheit	1686–1736	German	the first liquid-in-glass thermometer
Anders Celsius	1701–1744	Swedish	the Celsius or centigrade scale
James Watt	1736–1819	Scottish	improving the steam engine
Luigi Galvani	1737–1798	Italian	studying 'animal electricity'
Alessandro Volta	1745–1827	Italian	the first electric cell or 'battery'
André Ampère	1775–1836	French	studying electric currents
Hans Oersted	1777–1851	Danish	discovering the magnetism of an electric current
Georg Ohm	1789–1854	German	the law of electric circuits
Michael Faraday	1791–1867	English	the dynamo, electromagnets, the idea of 'fields'
James Joule	1818–1889	English	experiments with heat
James Dewar	1842–1923	Scottish	the dewar, or vacuum flask
Wilhelm Röntgen	1845–1923	German	discovering X-rays
Thomas Edison	1847–1931	American	the gramophone, the light bulb
Henri Becquerel	1852–1908	French	discovering radioactivity
Sir J. J. Thomson	1856–1940	English	discovering the electron
Heinrich Hertz	1857–1894	German	showing electromagnetic waves
Pierre Curie	1859–1906	French	experiments with radioactivity
Marie Curie	1867–1934	Polish/French	separating polonium and radium
Ernest Rutherford	1871–1937	New Zealand	founding nuclear physics
Guglielmo Marconi	1874–1937	Italian	inventing radio
Albert Einstein	1879–1955	German/American	many new theories about light, movement, energy, and mass
John Logie Baird	1888–1946	Scottish	making television
Enrico Fermi	1901–1954	Italian/American	the first atomic reactor
Robert van de Graaff	1901–1967	American	the high-voltage generator

Measuring

As a physicist, you should: OBSERVE—*watch closely to see what is happening;* DESCRIBE—*write down and record what you observe;* EXPERIMENT—*try out different things to see what happens; and* MEASURE. *This unit is about measuring.*

Estimating and measuring

Estimating Very often people make *guesses*. You can guess how far it is from London to Bristol or how warm a room is. If you think carefully about your guess and use all your knowledge and common sense, you are making an *estimate*. Estimates are often very useful, but they're not very exact and can sometimes be very wrong.

Sometimes your senses can mislead you. Which line in Figure 1 is longer? Check by measuring with a ruler. Your sense of touch can also play tricks. Try the experiment in Figure 2.

Using instruments Most of the measurements you make in physics need to be fairly accurate. Instead of trusting your senses you can use measuring instruments called *meters*. Every meter measures something, e.g. a thermometer measures how hot or cold something is. Every meter has a *scale* marked on it. The thermometer in Figure 3 is marked with a Celsius scale. This thermometer measures temperature in degrees Celsius. Degrees Celsius are called *units of measurement*. Every meter in physics has a scale, and every scale is marked in certain units.

Units for measuring in Temperature is measured in degrees Celsius (°C) or in Kelvin (K). 0°C = 273 K. Time is measured in seconds. Length is often measured in metres. Everything in physics is measured in its own special units. The table on page 7 shows some of the quantities that are measured, and the standard units they are measured in. These standard units are called *SI units*, and they will be used all through this book. (SI comes from the French words for 'international system'.)

Area and volume

Area It is often useful to know the area of something – perhaps a carpet, a table, or a field. You can work out the area of something using a simple rule:

> area = length × breadth

The area of a girl's thumbnail is about one centimetre squared (1 cm²). An area of one metre squared (1 m²) is much larger. Think of a square table top that is about as long as a man's arm.

Volume It is also useful to know how much space something takes up. This is called its *volume*.

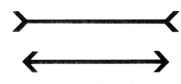

Figure 1 Which line is longer?

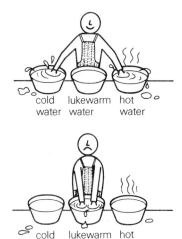

Figure 2 Put one hand in a bowl of cold water, the other in a bowl of hot water. Count to about 40, then put both hands in a bowl of lukewarm water. To one hand the water seems cold, to the other it seems quite warm.

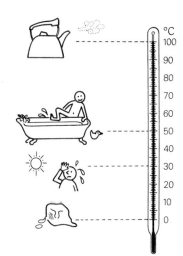

Figure 3 This thermometer measures temperature in degrees Celsius.

5

You can work out the volume of something using this rule:

volume = length × breadth × height

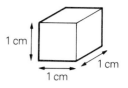

Figure 4 The volume of the cube is
1 cm × 1 cm × 1 cm = 1 cm³.

Small volumes, such as the cube in Figure 4, or the inside of a car or motor-bike engine, are usually measured in centimetres cubed (cm³). Larger volumes are measured in metres cubed (m³).

Volume of liquid   You often need to know how much space a liquid takes up. Figure 5 shows a special container for measuring the volume of a liquid. It is called a *measuring cylinder*. When you look for the water level in a measuring cylinder you will see that its surface is not flat, but curved. This curve is called the *meniscus*. Always read the scale where it meets the *bottom* of this curve, with your eye at the *same* level. The correct reading in Figure 5 is 180 cm³.

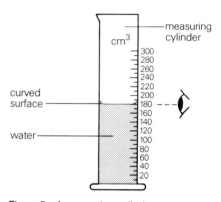

Figure 5   A measuring cylinder measures the volume of a liquid.

Force, mass, and weight

Force   A force is a push or a pull. Forces are measured in *newtons*, N for short. A force of 1 newton (1 N) is extremely small.

Mass   When you buy food you often buy it in *grams* (g) or *kilograms* (kg). The number of grams tells you the *mass* of the food. The mass of an object is a measure of how much material there is in it. A 2 kg bag of sugar contains twice as much sugar as a 1 kg bag.

Weight   Every object on Earth has some mass, but it also has a certain *weight*. Weight is a special kind of force. The weight of an object is the pull of gravity on it. The force of gravity pulling on a packet of sugar is its weight. Like any other force, weight is measured in newtons. A 2 kg *mass* of sugar has a *weight* of about 20 N. It would weigh less on the moon, because the force of the moon's gravity is weaker (about one-sixth of the Earth's gravity).

Remember that:

■ Mass is the amount of material in an object. Mass does not change. Mass is measured in grams or kilograms.

■ Weight is a force. It is the pull of gravity on an object. It can change as gravity changes. Weight is measured in newtons.

Density

Which has the most mass, a kilogram of feathers or a kilogram of coal? The answer is that they both have the same mass. But the coal will take up much less space than the feathers. The mass of the coal is much more *densely* packed. Coal has a *high density*, feathers have a *low density*.

Comparing density   Figure 6 shows one centimetre cube of some different materials. They all have the same volume but they have different masses. The number of grams in a centimetre cube, g/cm³, is a measure of the *density* of the material. The mass of 1 cm³ of lead is about 11 g. It is more dense than iron (8 g/cm³) but less dense than gold (19 g/cm³).

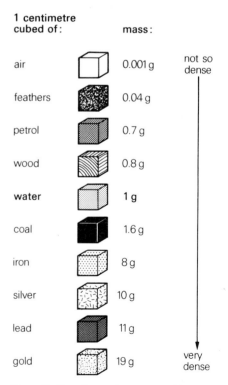

1 centimetre cubed of:		mass:	
air		0.001 g	not so dense
feathers		0.04 g	
petrol		0.7 g	
wood		0.8 g	
water		**1 g**	
coal		1.6 g	
iron		8 g	
silver		10 g	
lead		11 g	
gold		19 g	very dense

Figure 6   Density can be compared by comparing the same volume of different materials.

You can also compare densities by comparing the mass of 1 m³ of each material. The mass will then be much larger and will be measured in kilograms.

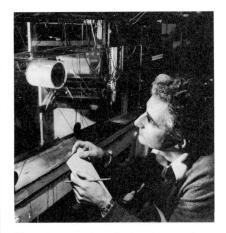

Summary

1 Estimates, or sensible guesses, are a useful part of science. But beware of your senses – they can mislead you.

2 Measuring instruments are called meters. Every meter has a scale, and measures in certain units. SI units are used in physics.

3 Mass is the amount of material in a substance. Weight is one type of force and is measured in newtons.

4 Some materials are very dense – their mass is packed into quite a small volume. The density of different materials can be compared by comparing the masses of 1 m³ or 1 cm³ of each one; in other words, by comparing the same volume of each material.

Observing, experimenting, and measuring are important activities in science. This scientist is investigating ways of using energy from waves in the sea to provide electricity.

quantity to be measured	units	for short	other information
mass	kilogram	kg	1 kg = 1000 g
	gram	g	
length	metre	m	1 m = 100 cm = 1000 mm
	centimetre	cm	1 cm = 10 mm
	millimetre	mm	
time	second	s	
area	metre squared	m²	1 m² = 1 m × 1 m
	centimetre squared	cm²	
volume	metre cubed	m³	1 m³ = 1 m × 1 m × 1 m
	centimetre cubed	cm³	
speed (and velocity)	metre per second	m/s	
acceleration	metre per second squared	m/s²	
force (including weight)	newton	N	
pressure	pascal	Pa	
density	kilogram per metre cubed	kg/m³	
	gram per centimetre cubed	g/cm³	
energy	joule	J	
	kilojoule	kJ	1 kJ = 1000 J
power	watt	W	
	kilowatt	kW	1 kW – 1000 W
electric current	ampere	A	
voltage	volt	V	

This table shows the standard units, called SI units, that quantities are measured in.

Is seeing really believing?

'It's not what you see, it's the way that you see it'

Two famous sayings: 'The camera never lies' and 'Seeing is believing'. Are they true?

Guessing the length Look at the lines marked **A** and **B**. Which line is longer, **A** or **B**? Now measure each one with a ruler. Do you still trust your eyes? Is this hat as broad as it is long?

Now look at the two squares below. Which square covers the larger area? When it comes to judging length and area your eyes can deceive you. It is much safer to use an instrument. That's what rulers are for.

Now you see it, now you don't 'There are two sides to every story.' There are also two ways of looking at a picture or a drawing. This drawing can be seen as two faces about to kiss each other. But if you concentrate hard on the white, it looks like a wine glass. . . .

Look at these simple drawings. Are you looking at the underside of the box, or at the three inside walls? It depends which side you stare at most . . . Are you looking at the inside or the outside of the open book? It keeps changing.

Scientists often explain things in two different ways, depending upon the way they see things. You will come across some examples later.

Seeing the wood from the trees You cannot always see things clearly because of their surroundings.

Look at these two long straight lines. Are they parallel? They don't look it, but they are the same distance apart all the way along.

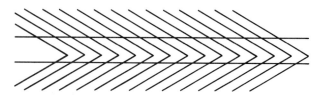

These two circles are both the same size. Check them with a ruler. But one looks smaller because the arrows direct your eyes inwards. The other looks larger because your attention is directed outwards.

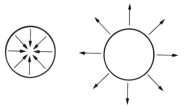

This optical illusion is nearly 100 years old. The drawing makes one figure look further away from you than the others. You expect him to be larger. But all three figures are the same size.

Whenever you *observe* in science try to ignore the messages or signals that don't really matter.

Even the camera can lie These women are both the same size. But one looks like a giant compared with the other! Do you know the reason?

Quite simple – you expect the room to be an ordinary square shape. In fact the room is shaped like this:

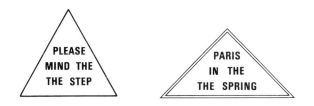

The woman that looks *half* the size of the 'giant' is *twice* as far away from the camera.

Never take things for granted. People often see what they *expect* to see. People often read what they expect to read. Read the message in each triangle. Then read each word again out loud.

PLEASE
MIND THE
THE STEP

PARIS
IN THE
THE SPRING

Learning to 'see' properly Very interesting, you might say, but what have these optical illusions got to do with physics? There are three important lessons to be learnt from them:

■ the human eye, and the camera, can lie. The brain can be tricked. This is one reason for using instruments.

■ people often see what they want or expect to see. Some ancient astronomers believed that the Earth was at the centre of the Universe. Every time they saw something that made them doubt this they ignored it. They were guilty of 'scientific dishonesty'.

■ you need to look out for the right things. Some messages or clues are important. Some are not important. The great discoveries in science have been made by people who noticed the important things, as you will see later.

Scientists have to learn how to *observe*. Learning science involves learning how to make *careful*, *honest*, and *well-trained* observations. Finally, just to confuse you, see if your eyes can make sense of these impossible objects. . . .

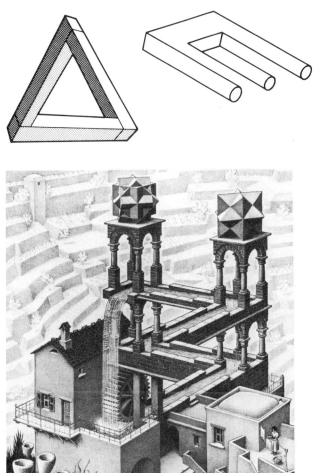

9

Introduction Exercises

1 Estimate, in centimetres, the length of:
   a) your finger
   b) your foot
   c) your body.
   Then measure each length with a ruler to check your estimates. How close were they?

2 Estimate the temperature of:
   a) the room you are in
   b) your body
   c) a hot summer day.

3 What is a meter? Describe three meters that people use in everyday life.

4 How many millimetres are there in:
   a) 1 cm? b) 5 cm? c) 0.5 cm?
   d) 500 cm? e) 1 metre?

5 How many centimetres in:
   a) 0.5 m? b) $1\frac{1}{2}$ m? c) 3.6 m?
   d) 1 km?

6 How many cubic metres of water would it take to fill this swimming pool? Suppose it was emptied using a bucket with a volume of $\frac{1}{4}$ m³. How many bucketfuls would be needed?

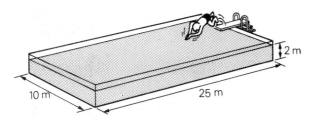

10 m   25 m   2 m

7 Explain the difference between mass and weight. What unit is each one measured in?

8 What does the word 'dense' mean? Write down three materials that are dense, and three that are not so dense.

Things to do

Find the 18 famous scientists in this 'Wordsearch'. Copy the wordsearch out before you mark in the answers. The words can go across, up, down, diagonally, but only from left to right. Here are some clues to help you:

■ three of them were Ancient Greeks
■ one used a leaning tower
■ an Englishman, fond of apples?
■ 'ome sweet 'ome
■ four of the scientists had plenty of shocks
■ his first name was Albert
■ his work began in earnest in 1911
■ invented the nuclear reactor
■ two were very interested in planets
■ this Italian seemed to like frogs
■ 'what' a pair of 'jewels'

```
R X F R F E R M I J H A A E I I J H J C
T W L U S E R H G T T W H T W B N F C J
T F M T D F K K C L W O P V M V G E J F
T A T H T N P K O N D B A F X H F I I G
V R U E G G D V P A O B H H C H T N X A
F A H R W S F Z E R T S W E N C J S N L
E D E F F I C S R I P U S A Z H L T E V
X A H O S K F P N S A M P E R E L E W A
M Y P R G H D T I T R O N P G P V I T N
T P J D V G D H C O K E L B T P J N O I
L G O U E L Y D U T P R W A T T O W N O
V A U H F M S L S L O S S S Q J T N S B
J D L H M S O J M E A T U D I W H K X N
X N E K U M I C F S A E G A L I L E O X
N X F Z E X Y S R T F D E O O E G P P N
E M G H L A R C H I M E D E S N I L D F
E F D Q K Q Y V J T T Z U L F I N E D A
S W G T O S Q H B U K U A B W X T R Z P
O S C N B Q G E F O B B S E D T O B H H
Z I M H R P O C U R X T N N E X X G P Y
```

TOPIC 1

Matter

Many solids are made up of crystals. Under a polarising microscope, crystals of
sugar cane look like flower petals.

1.1 Solids, liquids, and gases

What's the difference between ice, water, and steam? Why do all balloons leak? How can some animals walk on water? What things can be compressed and stretched? The answers to these questions depend upon the three types of material and the way they behave.

Materials

All the things you see around you are made up of some material. A wall might be made of brick or stone, a lake is full of water, the air you breathe is full of different gases. Brick, stone, water, and air are all different types of *material*.

Look at the car in Figure 1. You can see that all the materials in and around the car can be put into one of the three boxes: either *solid, liquid*, or *gas*. The same is true of *all* the materials on Earth. Some examples are shown in Figure 2.

solids	liquids	gases
brick	water	steam
stone	lemonade	'fizz' (carbon dioxide)
lead	petrol	oxygen
coal	oil	air
wood	beer	helium
salt	vinegar	hydrogen

Figure 2 Some examples of solids, liquids, and gases

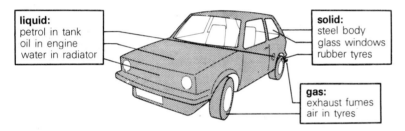

liquid:
petrol in tank
oil in engine
water in radiator

solid:
steel body
glass windows
rubber tyres

gas:
exhaust fumes
air in tyres

Figure 1 A car consists of solids, liquids, and gases.

Figure 3 A gas has to be kept in a container.

Gases

Think of exhaust fumes, the smell of perfume, and the gas from a cooker. They are all gases. Gases like these are hard to keep in one place. They have no fixed shape or volume. They have to be bottled up very tightly and trapped on all sides or they will escape. To keep a gas you must put it in a *container*, as shown in Figure 3.

A gas will not stay in one part of a container but will spread out to fill it, even if it is as large as a room. Think of strong smells, such as the smell of petrol, the fragrance of a flower or perfume, bread baking, or the warning smell of leaking gas. The way that these smells spread through the air or across a room is called *diffusion*. Gases always spread out to take up as much space as they can: they *diffuse*. Even a balloon that is very tightly tied goes down after a few days. The air inside it leaks out, or diffuses, through the walls of the balloon.

Finally, gases can be squeezed or *compressed* into a smaller and smaller space. For example, the air in car tyres has been squeezed in. Figure 4 shows how the air in a bicycle pump can be compressed.

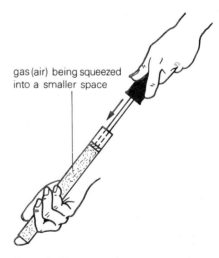

gas (air) being squeezed into a smaller space

Figure 4 Gases can be compressed.

Liquids

Think of petrol, water, and oil. They are all liquids. Like gases, liquids have to be kept in containers. But, as Figure 5 shows, one side can be left open and the liquid will not escape. A liquid can alter its shape, but not its volume. The open side of a liquid is called its *surface*. The liquids in Figure 5 all have the same shape as their container, with a flat surface at the top.

Funny things happen on the surface of liquids. For example, some insects can walk on water, as Figure 6 shows. You can even float a steel needle on water. Figure 7 shows you how: place the needle on a piece of tissue paper, then lay it gently on the surface. The tissue becomes wet and sinks but the needle stays afloat. It is as if water has a kind of 'skin' on its surface. This is called its *surface tension*.

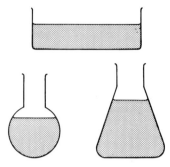

Figure 5 A liquid takes the shape of its container.

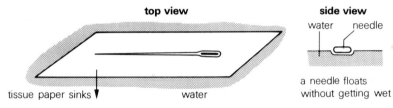

Figure 7 A steel needle will float on water.

Most liquids can be 'soaked up'. A piece of blotting paper or a sponge can soak up ink or water. Houses often have a special waterproof barrier between the layers of brick near the ground to stop the dampness from rising, as Figure 8 shows.

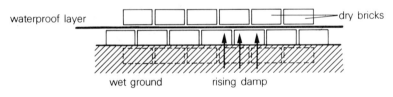

Figure 8(a) A damp-proof course works because the waterproof layer blocks the rising damp.

Like gases, liquids can diffuse. If you place one drop of ink in a beaker of water, the colour of the ink gradually spreads all through the water. This is called *liquid diffusion*.

Solids

Solids don't need to be kept in a container. They have a fixed shape and volume, as Figure 9 shows. Unless they are melted down, cut, or crushed they will keep this shape.

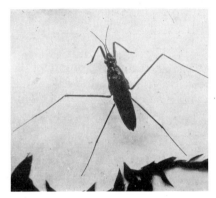

Figure 6 This pondskater is walking on water.

Figure 8(b) A waterproof layer between the bricks will stop rising damp.

cricket ball

brick

telephone

car body

Figure 9 Solids have their own shape.

Have you ever looked closely at a grain of sugar or a pinch of salt? Sugar and salt are both made up of tiny *crystals*. Many solids are found in the form of crystals. Each crystal has its own shape, sometimes very beautiful, as Figure 10 shows. Snowflakes have thousands of different patterns.

Many solids can be stretched. A rubber band, a piece of elastic, a guitar string all become longer when they are pulled. A piece of steel or copper wire can be wrapped round a pencil to make a spring. This spring can be stretched or squeezed, as Figure 11 shows. Let go of the spring and it 'springs' back to its first shape.

stretched

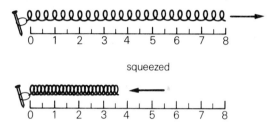

squeezed

Figure 11   A spring can be stretched and squeezed.

Figure 10   The top photo shows a snowflake crystal. The bottom photo is a crystal of sugar magnified 160 times.

Summary

1   Materials can be labelled as solid, liquid, or gas.

2   A gas has no fixed shape or volume. It takes the shape of its container.

3   A gas spreads out to fill its container by a process called diffusion.

4   A gas can be compressed into a small space.

5   A liquid has a fixed volume but no fixed shape. It takes the shape of its container.

6   A liquid seems to have a skin on its surface. This is called its surface tension.

7   Liquids can diffuse and most liquids can be 'soaked up'.

8   A solid has a fixed shape and a fixed volume.

9   Many solids exist as crystals.

10   Many solids can be stretched and squeezed.

Exercises

1   Make a table showing each material as a solid, liquid, or gas:
    stone, diamond, flour, sea-water, air, vinegar, wine, hydrogen, brick, helium, treacle, butter, ice, custard, paint, plastic, oxygen, steam.

2   Describe two examples which show that liquids have surface tension.

3   Even a tightly tied balloon goes down after a few days. Explain why.

4   Give three examples of liquid diffusion.

5   Why do houses have a waterproof barrier between layers of brick close to the ground? What is this barrier called?

6   Copy out and complete the statement:
    a) Gases have no fixed _____ or _____ .
    b) The spread of a gas throughout a room is called _____ .
    c) Insects can float on water because of _____ _____ .
    d) Solids normally have a fixed _____ and _____ .

1.2 The Particle Theory

Scientists have come up with a theory to explain why solids, liquids, and gases behave the way they do. The theory is so important it has a special name. It is called the Particle Theory.

The Particle Theory says that *all* materials are made up of very tiny bits or pieces called *particles*. These particles are so small they cannot be seen with even the strongest microscope. But scientists believe that they are there. The particles are always *moving*. As a material gets hotter its particles move around faster and faster.

The particles in a material **attract** each other. In solids the particles attract each other strongly, in liquids the attraction is much weaker, while in gases the particles hardly attract each other at all. Figure 1 shows how scientists imagine the particles inside a solid, a liquid, and a gas.

These particles are too small to be seen. So how do scientists know that they are there, and that they are moving? Why do scientists believe in this theory?

Support for the theory

The best evidence or clue was first noticed by a scientist called Robert Brown in 1827. He saw pollen grains floating on water and noticed that they were moving around in a zigzag, haphazard way like a drunken man. He could see the pollen grains through his microscope but *not* the tiny particles inside the water. These water particles were moving around and bumping into the pollen, making the grains jump around.

Nowadays this movement is called *Brownian motion*. Figure 2 shows how you can see it. A microscope is used to look into a small glass box full of smoke (a 'smoke cell'). The smoke particles move about in a jumpy, jerky way in all directions. They are being pushed and shoved around by the moving air particles.

Brownian motion is the scientist's best clue for the theory that all materials are made up of moving particles. But what *are* these tiny particles of matter?

The particles

Scientists call the tiny particles *atoms*, a name first used by an ancient Greek, Democritus, and meaning 'unsplittable'. Atoms are like the letters of the alphabet. Just as letters can be made into words, atoms nearly always join together to make *molecules*. Atoms and molecules are the building bricks of all materials or matter.

How big are molecules? Look at this dot of ink ● It contains millions and millions of ink particles or molecules. It is impossible to imagine the size of *one* molecule – it is less than a thousand-millionth of a metre. If you could lay a thousand million particles of matter, or molecules, side by side they wouldn't even stretch for a metre.

From now on these tiny particles will be called *molecules*.

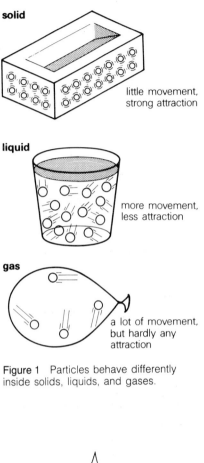

Figure 1 Particles behave differently inside solids, liquids, and gases.

solid — little movement, strong attraction

liquid — more movement, less attraction

gas — a lot of movement, but hardly any attraction

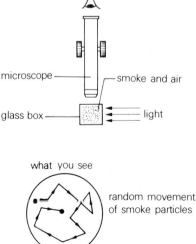

microscope — smoke and air
glass box — light
what you see — random movement of smoke particles

Figure 2 You can see Brownian motion under a microscope.

Using the theory

To help you remember the Particle Theory here are the main ideas again:
- everything is made up of tiny particles
- these particles are always moving
- the particles attract each other.

The main reason that scientists believe in the Particle Theory is because it is so useful for explaining things. Here are some examples:

Melting and boiling   Figure 3 shows water as a solid, a liquid, and a gas. In the ice lolly water molecules are moving quite slowly. They are attracted together and held in a lolly shape. As the ice lolly starts to warm up the molecules move faster – they become more *excited*. When the ice lolly melts the molecules break away from each other. Solid ice then becomes liquid water. Finally, when water starts to boil some of its molecules break away from the other ones attracting them in. Water turns to steam and the molecules of gas spread out into the air.

Stretching and squeezing   The Particle Theory can also explain why metal springs can be stretched and squeezed. When you stretch a spring the molecules inside it are pulled further apart. But they still attract each other. So when you let the spring go it springs back as the molecules pull each other together again.

Molecules like to be close to each other – but not too close. When you squeeze a spring, or a dry sponge, it goes back into shape when you let go. Molecules are very fussy – they like to be just the right distance apart.

The particles inside a crystal are arranged in special patterns. They attract each other and hold the crystal in this shape. Figure 4 shows how you can imagine the particles inside a crystal.

Gases   The molecules in a gas have a lot of space between them. This means a gas can be compressed. When a gas is squeezed or compressed the molecules are pushed much closer together and the gas takes up less room. Some gases, if they are squeezed hard enough, into a very small space, may change into a liquid.

The molecules in a gas move around very quickly. They do not hold each other together in a definite shape. This is why gases have to be kept in containers. The fast-moving molecules spread out or diffuse to take up as much space as they can. This is why smells travel so quickly.

Liquids   The molecules inside a liquid do not move as freely as they do in a gas. But they move around much more freely in a liquid than in a solid. These moving molecules help to explain why water can be soaked up by tissue paper or blotting paper. The moving water molecules travel up into the tiny air spaces inside the tissue, as Figure 5 shows.

The theory also explains why the colour of an ink drop spreads through water. The liquid molecules are continually moving, in every direction. As they move and bump into each other, the ink molecules spread out or diffuse to fill all the available space.

solid water (ice)

liquid water

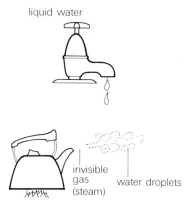
invisible gas (steam)
water droplets

Figure 3   Water can exist as a solid, liquid, or gas.

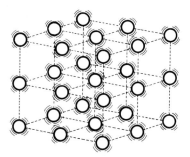

Figure 4   Particles inside a crystal are arranged in a special way.

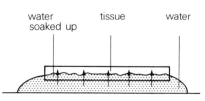
water soaked up   tissue   water

Figure 5   Moving water molecules can be soaked up by a tissue.

Surface tension can also be explained by the Particle Theory. The molecules at the surface of water are pulled *in* by the attraction of the molecules lower down in the liquid. This force pulling them in makes the surface of water seem like a stretched skin.

Some strange materials

You now know an important theory that scientists have, and how it can be used. You know that all materials are either solid, liquid, or gas. But are they? *Most* materials fit neatly under one of these headings. Some do not. What about jelly, cold custard, or non-drip paint? Are they solids or liquids? It is hard to decide.

One funny material that you can buy from joke shops is called 'potty putty'. It stretches, moulds like clay, can be rolled into a ball and bounced. Like some solids it can be broken with a hammer. But if it is left on a table it gradually collapses into a puddle of putty. Is it a solid or a liquid?

The colour of an ink drop spreads through water as the ink molecules move about.

Summary

1   The Particle Theory is used by scientists to explain why solids, liquids, and gases behave the way they do.

2   The Particle Theory says that everything is made up of tiny particles which are always moving and attracting each other.

3   Brownian motion is the random zigzag movement of small objects like smoke particles. The Particle Theory explains this movement as moving air particles bumping into the smoke particles.

4   The very tiny particles are called atoms. Atoms join together to form other very tiny particles called molecules.

5   Molecules in a solid are strongly attracted to each other and do not move about freely. This gives the solid its fixed shape and volume.

6   Molecules in a liquid are not as strongly attracted as in a solid and can move around more freely. This means the liquid can spread out into the shape of its container.

7   Molecules in a gas are not very strongly attracted and can move around very freely. This is why a gas has no fixed shape or volume.

8   The arrangement of molecules in solids, liquids, and gases explains why melting, boiling, stretching, and squeezing can occur.

Brownian motion is named after Robert Brown.

Exercises

1   Write down three important facts about gases, three about liquids, and three about solids.
    What theory is used to explain these facts?
    Write down the main ideas of this theory.

2   Every substance is made up of tiny particles. What are these particles sometimes called? What evidence is there that they are always moving?

3   Why are gases easier to squeeze than liquids or solids?

4   Explain what happens to the tiny particles when:
    a) water turns to steam
    b) a spring is stretched and then let go
    c) air is compressed.

5   In your own words explain why:
    a) gases need to be kept in a container
    b) a balloon gradually leaks even if it is tightly tied
    c) the colour from an ink drop spreads throughout a beaker of water.

1.3 Melting and boiling

The Particle Theory can explain why solids melt and why liquids boil. Water normally boils at 100°C at sea level . But if you lived on top of Mount Everest the water for your tea would boil at only 70°C.

Changing state

You have seen water in its three different states: solid (ice), liquid (water), and gas (steam). In fact all substances can exist as solid, liquid, or gas if they become hot enough or cold enough. Oxygen becomes a liquid at −183°C. It even becomes solid at −219°C.

As things get colder and colder their molecules move more and more slowly. Eventually, at the coldest temperature possible, the molecules hardly move at all. This temperature is called *Absolute Zero*, and is −273°C. Nothing has yet been cooled this far, though physicists are still trying! On the Kelvin scale, −273°C = 0 K.

Iron becomes a liquid at 1535°C, and will even boil at 3027°C. Imagine iron as a gas. It would be called 'iron vapour'.

The table below shows some interesting melting and boiling points.

The temperature inside the Earth is so hot that rocks are liquid. This photo shows liquid rock, or lava, flowing from the side of a volcano into the sea.

material	melts at		boils at	
hydrogen	−259°C	(14 K)	−253°C	(20 K)
oxygen	−219°C	(54 K)	−183°C	(90 K)
mercury	−39°C	(234 K)	357°C	(630 K)
water	0°C	(273 K)	100°C	(373 K)
lead	327°C	(600 K)	1744°C	(2017 K)
copper	1083°C	(1356 K)	2595°C	(2868 K)
iron	1535°C	(1808 K)	3027°C	(3300 K)

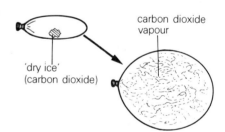

Figure 1 Solid carbon dioxide can be used to blow up a balloon.

Some materials can change straight from a solid to a gas. Carbon dioxide is one example. It misses out the liquid stage. Figure 1 shows how a piece of solid carbon dioxide, or 'dry ice', can be used to blow up a balloon. As the solid becomes a gas it takes up much more space (about 600 times more) and inflates the balloon.

Hidden heat

Look at the simple experiment shown in Figure 2. As the ice is melting its temperature stays at 0°C until *all* the ice has changed to water. If you carry on heating the beaker the water gradually becomes warmer and warmer until it starts to boil at 100°C. What happens to its temperature now? If the water is clean or *pure* it stays at 100°C until *all* the water has changed to steam.

You can draw a simple graph to show how the temperature of the ice changes.

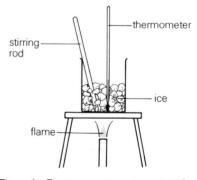

Figure 2 The temperature stays at 0°C until all the ice has melted.

Figure 3 shows how the temperature 'stands still' or stops rising when the ice is melting, and when the water is boiling.

What is happening to the heat of the flame when the graph stands still? The Particle Theory can explain the graph. The particles inside the ice hold each other together with strong forces. When the ice is melting all the heat is being used to pull the particles or molecules of ice away from each other. The ice becomes water. It melts.

The heat used to melt the ice is called hidden or *latent* heat. Latent heat is also needed to make the water boil. The water molecules hold each other in with quite strong forces. As they are heated they jump around faster and faster, become more excited, and finally escape from the water. These escaping molecules are steam.

The hidden heat needed to make a solid melt or a liquid boil is called *latent heat*.

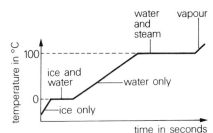

Figure 3 This graph shows how the temperature changes as the ice melts to water and the water boils to steam.

Changing the melting point

Adding salt   Normally ice melts into water and water freezes into ice at 0 °C. This is called the *melting point* or freezing point of water. But it can change.

Have you ever seen workmen throwing salt and grit on the roads in winter? The salt is used to stop water freezing on the roads. When salt is added to water it freezes at a temperature colder than 0 °C, sometimes −3 °C, −4 °C, or even lower. The salt on the roads *lowers* the freezing point of water. Unless the temperature drops well below 0 °C no ice is formed and the roads are safe. For the same reason the sea around Britain hardly ever freezes.

Pressure and melting   You can use a very thin metal wire, like 'cheesewire', to cut through a block of ice, as Figure 4 shows.

Salt lowers the freezing point of water, so the sea does not freeze unless the temperature drops well below zero. This photo was taken in Greenland.

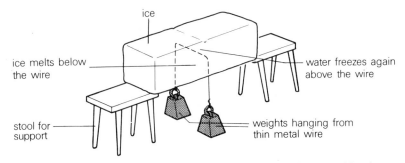

Figure 4   Pressure can be used to cut through a block of ice with a thin wire.

The pressure of the wire helps to melt the ice *under* it. As the ice melts the wire moves down. But the water *above* the wire soon freezes again, leaving a solid block. The wire cuts through the block of ice without breaking it.

The pressure of the cheesewire *lowers* the melting point of the ice below it. This makes it melt.

Pressure lowers the melting point of ice. In a glacier, pressure from the ice helps to make a slippery film of water for the glacier to slide on.

Changing the boiling point

Adding salt Water usually boils at 100°C, its *boiling point*. But, just like the melting point, this can change. Adding salt to water *raises* its boiling point. Instead of boiling at 100°C the water boils at 120°C or even higher. Potatoes cook more quickly in salted boiling water – they also taste better!

Pressure cooker Some people cook with a pressure cooker, as in Figure 5. Because the lid is completely airtight, pressure builds up inside the cooker. The extra pressure above the water stops the water molecules from escaping as steam at 100°C. They need to become more excited and move faster to force their way into the steam pressing down on them. The water does not boil until it reaches about 120°C. This is much hotter than in a saucepan and the food cooks much more quickly.

Lowering the boiling point If the Particle Theory is true water should boil more easily if there is less pressure on it. The molecules should find it easier to escape. This is exactly what happens. Water boils at 100°C at sea level. Higher, where the air is thinner, the boiling point gradually drops. In Mexico, water boils at about 95°C. Figure 6 shows how the boiling point gets lower as you go higher and higher. Have you tried making a good cup of tea on Mount Everest?

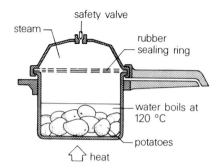

Figure 5 Water boils at a higher temperature when there is more pressure on it.

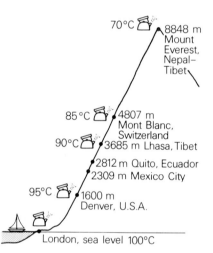

Figure 6 Water boils at a lower temperature when there is less pressure on it.

Summary

1 A solid melts when its molecules move fast enough to form a liquid. A liquid boils when its molecules escape to form a gas.

2 The heat needed to change a solid to liquid, or a liquid to gas, is called latent or hidden heat. This heat is needed to tear the molecules away from the forces holding them inside a solid or liquid.

3 The melting point of ice can be lowered by adding salt or by pressing on it with a fine wire.

4 The boiling point of water can be raised by adding salt or lowered by lowering the air pressure above it.

Exercises

1 What are the three different states of water? Using the Particle Theory, explain how ice melts to become water, and how water boils to become steam.

2 What is the usual boiling point of:
  a) oxygen?
  b) iron?
  c) water?
  d) lead?
  Write down the melting point of each one. What state is each one in at 20°C?

3 What is special about the temperature −273°C?

4 Explain what latent heat is.

5 Why is salt spread on the roads in winter?

6 Explain how a very thin metal wire can cut through a block of ice without breaking it.

7 Potatoes are cooked in salty water to make them taste nicer. How does the salt affect the boiling point of water?

8 Explain, using a drawing, how a pressure cooker works. Why is it useful?

9 Why do mountaineers find it so difficult to make a good cup of tea? (*Clue:* Hotter water makes better tea.)

1.4 Water, water everywhere

A fridge or a freezer actually warms up the room it is in. You would probably feel hotter in a tropical jungle than in the Sahara desert. How can physics explain these strange claims?

Evaporation and condensation

Boiling and evaporating   When water boils it changes quite quickly from a liquid into a gas (steam). This only happens at its boiling point, usually about 100°C. But water *can* change into a gas *without* boiling. Puddles of water, after a storm of rain, soon disappear. Where does the water go? The liquid, water, changes into gas and becomes part of the air around you. The gas is called *water vapour*.

Whenever a liquid changes to a gas without boiling we say that it *evaporates*.

Volatile liquids   Some liquids change into gas, or evaporate, very easily. Petrol, meths, perfume, and ether are all 'smelly' liquids. Particles or molecules of petrol leave the liquid and go into the air – these particles enter your nose and make a smell. Whenever you smell something tiny particles of it enter your nose. Liquids that evaporate easily are called *volatile*.

Condensation   Have you ever boiled a kettle near a window? Steam from the kettle changes back into water when it meets the window. The steam *condenses*. Whenever a gas or vapour changes back into a liquid, condensation is taking place.

Condensation is the opposite of evaporation. You can see condensation on a cold bath tap, a window, a car windscreen, the bathroom mirror, or on a boy's spectacles when he comes indoors on a cold day.

Clouds and rain are made by condensation.

Latent heat

Keeping cool   Try putting a few drops of meths or petrol on the back of your hand. As the meths dries up or evaporates your hand feels cool. The sweat on your body does the same thing – it helps to keep you cool. The meths, or the sweat, takes *latent heat* away from your body so that the liquid can change into gas, or evaporate.

Whenever a liquid evaporates it takes in heat from objects near it. This is why you feel cold when you get out of a swimming pool.

Ether   Figure 1 shows how ether uses latent heat to evaporate. Ether is a rather dangerous, volatile liquid. As air is blown through it, ether evaporates and takes heat away from the water under it. Soon the water turns to ice. The ether has stolen latent heat from it – the liquid ether *uses* it to become gas, the water *loses* it to become ice.

Clouds are made by condensation of water droplets.

Water droplets often condense from warm air onto a cold window.

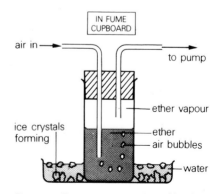
Figure 1   Ether evaporates by taking heat from the water.

Scalding steam Exactly the opposite happens when a gas, like steam, condenses. If steam meets your skin it gives up or loses heat to your body as it condenses into water. This extra heat means that steam makes a worse scald than boiling water. Whenever a gas condenses it gives up latent heat. This is how a refrigerator works.

Refrigerators Usually an electric motor pumps a liquid called freon around in a circle, as Figure 2 shows. This liquid is very volatile. Inside the fridge the freon evaporates and *takes in* latent heat. It steals heat from the inside of the fridge, making it cool. Outside the fridge the freon, which is now a vapour, is forced to condense as it is squeezed or compressed by the pump.

As the freon vapour condenses it *gives up* its latent heat. This is why the pipes at the back of a fridge feel warm. The freon is continually taking heat out of the fridge and putting it into the room.

Water in the air

Drying the washing On a warm, sunny, windy day water dries out or evaporates quickly. Water also evaporates quickly when it is 'spread out' to cover a large area. Hanging washing on a line exposes more of it to the air. You can use the theory that water is made up of tiny particles or molecules to explain why washing on a line dries faster on a warm, windy day. Water evaporates when some of its molecules break away or escape from the liquid into the air. If the water is warm the molecules move around more quickly and more escape into the air. If a large area of water is exposed the molecules have more chance of escaping. In a wind the water molecules which have evaporated are blown away leaving room for more to escape or evaporate into the air.

Humidity On some days you complain of being 'hot and sticky'. The sweat on your body does not evaporate and keep you cool. The air is already full of water vapour – it just can't take any more. The air is *humid*, which means it is very damp. The air in a jungle is humid or full of water vapour. In the Sahara desert the air is almost dry – it contains no water vapour. Any water will evaporate into the air straightaway.

Dew and mist Warm air can hold more water vapour than cold air. If warm air full of water vapour is cooled, some of the vapour changes or condenses back to water. On a cold night water often condenses on grass or the roofs of cars. This water is called *dew*. Water vapour in the air condenses on cold things, just as the steam in a bathroom condenses on a cold tap.

If the air is very damp and then suddenly cools you can actually see this water vapour as tiny droplets hanging in the air. This is *mist* or fog.

Clouds Clouds are made in the same way. There is always some water vapour in the air but usually it is invisible. As the air rises it cools down until the water vapour inside it changes back into tiny droplets of liquid water. These droplets of water collect together to make *clouds*.

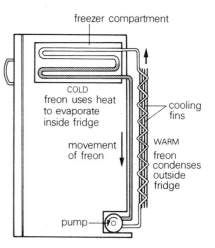

Figure 2 A fridge is cold inside because the liquid freon takes heat away from the fridge and uses it to evaporate.

The pipes at the back of a fridge feel warm because the freon vapour gives up latent heat as it condenses.

Fog is tiny drops of water vapour hanging in the air. In thick fog most of New York's skyscrapers are invisible from the ground.

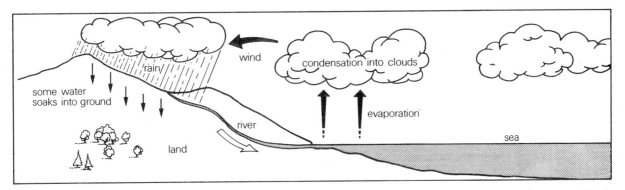

Figure 3 Water never disappears, it just changes its form and goes round and round in a cycle.

The water cycle

Figure 3 shows the water cycle. Water on the Earth's surface – in seas, lakes, rivers, and puddles – is changed into vapour by heat from the sun. It evaporates. As the damp air rises, over a hill or a mountain, it cools down until some of the water vapour condenses to make a cloud. If the air rises high enough the water may even freeze to make snow or hail. In a 'rain cloud' the water droplets collect together until they are so heavy they fall as rain. The rain fills the rivers that flow into the sea or lakes and the whole cycle begins again.

Summary

1   Liquids can change into a gas or vapour without boiling: they evaporate. When a gas or vapour changes back to a liquid it condenses.

2   When a liquid evaporates it takes in latent heat. When a gas condenses it gives up latent heat.

3   Inside a refrigerator a liquid called freon evaporates and takes heat away from the fridge. It condenses outside the fridge and gives up heat to the room.

4   The air often contains a lot of water vapour but it is usually invisible. Very damp air is called humid.

5   When air cools down the water vapour in it sometimes condenses to make fog, mist, or dew. Condensed water vapour in the sky forms clouds and often falls back to the Earth as rain.

6   The continual circulation of water is called the water cycle.

The continual circulation of water is called the water cycle. Water evaporates from lakes and rivers, forms clouds, and falls back to Earth as rain.

Exercises

1   What is the difference between evaporation and boiling?

2   What is the difference between evaporation and condensation?

3   What is a volatile liquid? Give four examples.

4   How does sweat cool your body?
    Why do people feel 'sticky' on humid days?

5   Explain why steam makes a worse scald than boiling water.

6   Describe how a fridge works, explaining how it takes heat in on the inside and gives heat out at the back of the fridge.

7   What kind of weather is best for drying washing? Explain why.

8   Explain the difference between dew and mist.

9   How are clouds formed?

10   Draw the water cycle and explain how it works.

1.5 Heat and matter

This Unit describes some of the ways that matter behaves when it is heated. Things grow larger, or expand. Heated air or water rises. Heat travels through a material. Some materials become red-hot and glow. You will learn the difference between expansion, convection, conduction, and radiation.

Solids expanding

Showing expansion When solids are heated they grow larger or *expand*. Figure 1 shows a way of proving this. The metal ball slips through the ring when it is cold. Try heating the ball – it expands, and becomes too big to slip through the ring. When the ball cools it shrinks back to its first size – it *contracts*.

The force of expansion If you heat a solid and *stop* it from expanding it pushes with a very strong force. This is why small gaps are left on concrete roads and at the ends of bridges, to allow for expansion when it is hot. Some bridges are even laid on rollers so that they can expand in summer and contract in winter, as Figure 2 shows.

Breaking a metal pin You will also get strong forces if you heat a metal bar and then stop it from contracting or shrinking to its first size. Figure 3 shows how a metal bar can be used to break an iron pin. First, the bar is heated and then the big nut is tightened so the bar cannot contract freely. As the bar cools down it pulls strongly against the iron pin until the pin suddenly snaps.

Liquids and gases expanding

Different liquids Solids only expand by a very, very small amount. You cannot usually *see* them expand – you know they expand and contract because of the strong pushing and pulling forces. Liquid expansion is easier to observe. If different liquids are put in identical narrow tubes and then heated, the liquid levels gradually rise. Some liquids rise more than others. For example, paraffin and alcohol expand much more than water.

Burst pipes When water is cold it is different from other liquids. Most liquids expand when they are heated, and shrink or contract when they cool down. But water actually *expands* when it is cooled from 4°C down to 0°C, and then it freezes. Ice takes up more space than water. This is why water pipes can burst in winter.

Fish in water Because ice takes up more space than water, it is less dense than water. This means that ponds do not freeze solid, and fish can stay alive in winter. Ice floats on top of the pond, the warmer water at 4°C stays at the bottom. The fish can swim around underneath the ice.

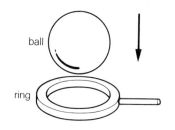

Figure 1 The ball goes through the ring when it is cold but not when it is hot.

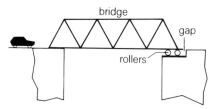

Figure 2 The rollers let the bridge move when it expands.

Gaps are left at the ends of bridges because the bridge expands when it gets hot.

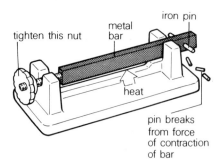

Figure 3 The force of contraction is strong enough to break the iron pin.

Gases expanding Liquids expand more than solids. Gases expand even more than liquids do – about 10 times as much. Figure 4 shows how you can demonstrate that air expands as it gets warmer. As your hands warm the air in the flask it expands and pushes the water in the tube down. Bubbles appear when there is no water left in the tube. Take your hands away and the air shrinks or contracts, the bubbles stop, and water goes back up the tube.

Using expansion

Air thermometers In about 200 BC a Greek called Philo used the fact that air expands so easily to make a rough thermometer. Galileo used a better one in 1592. It used the rise and fall of the water level in a tube to give a rough and ready idea of the air temperature. (See the photo on page 27.)

Liquid-in-glass thermometers Unfortunately the air is not very reliable. The atmosphere, or air around us, presses down on everything. This pressure, called *atmospheric pressure*, is always changing. Nowadays *liquid-in-glass* thermometers are used. The liquid in a small glass bulb expands inside a very narrow tube called a capillary tube. The more its temperature goes up the more the liquid expands. The glass tube has a scale on it, as in Figure 5, with two important marks called *fixed points*. These are 0°C (the temperature of pure melting ice) and 100°C (the boiling point of pure water). Dividing the gap between them into 100 equal spaces makes a *centigrade* or Celsius scale.

Mercury and alcohol thermometers The liquid mercury is often used because it expands *evenly* as it gets warmer, and can be used at high temperatures, up to 360°C. Unfortunately it freezes solid in cold places like the North Pole. Some thermometers use alcohol because it expands more than mercury and does not freeze until −115°C. It is also much cheaper, but it should not be used in boiling water.

Metals expanding The expansion of metals when they become hot is still used to make cartwheels. The cold metal rim is too small to go round the wheel. But after heating it expands and can be slipped on. As it cools it contracts and grips the wheel to make a tight fit.

Bi-metal strip Some metals, like brass, expand more than others. One special metal, called invar, hardly expands at all. A strip of brass and a strip of invar (or iron) can be riveted or welded together to make a *bi-metal strip*. When this is heated the strip bends as the brass expands, as shown in Figure 6. The brass takes the outside of the curve because it expands more than invar (or iron).

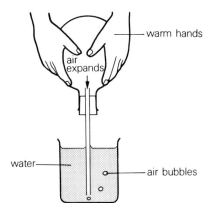

Figure 4 The air expands as it gets warmer and pushes on the water in the tube.

Philo made a rough thermometer in about 200 BC.

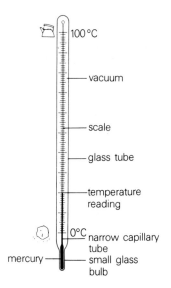

Figure 5 A thermometer uses expansion of a liquid to measure temperature.

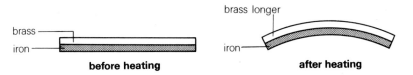

before heating   after heating

Figure 6 A bi-metal strip uses two metals which expand differently when heated.

Fire alarms Figure 7 shows how you can make a simple model of a fire alarm, using a bi-metal strip. In a fire the strip gets hot, bends to meet the contact, switches on the current, and the bell rings. Can you find out any other uses for bi-metal strips?

Why does matter expand?

All materials, solids, liquids, and gases, take up more space when they are hot. Brass, concrete, steel, water, mercury, alcohol, and air all expand if they are heated. Once again the Particle Theory can explain *why* this happens.

The theory says that everything is made up of tiny particles moving around in different directions. Imagine air held in a container by a piston which can move up and down, as in Figure 8. As the air is heated the particles or molecules move faster and faster, bumping into the piston, and pushing it back until they take up more space. The air expands.

The same thing happens in solids and liquids. When a metal bar is heated its molecules move around more quickly and take up more space.

Heat on the move: conduction

Heat can make a metal bar expand. It can also travel through metal and other materials. It always travels from hot parts to colder parts. Figure 9 shows a simple experiment to prove that heat travels through some materials better than others. The heat travels, or is *conducted*, along each rod until the wax melts and the nail drops off. The nail on the copper rod falls off first.

Some materials, especially metals like copper and silver, carry heat well. They are called good *conductors*. Heat travels through them as their molecules move around and bump into each other. The heat is passed from molecule to molecule, along the metal. This is called *conduction*.

Hot air rising: convection

When air is heated it takes up more space or expands. Have you tried holding your hand well above a candle flame or a burning match? You can feel the hot air rising from it. As the air near the flame gets hot it expands and becomes *less dense*. The hotter, less dense air rises and colder, denser air moves in to take its place.

Figure 10 shows how rising air can quickly warm a room. The hot air rises and the cooler air sinks to take its place – this movement of air is called *convection*.

Convection in water Convection also occurs in water. Hot water expands, takes up more space, and rises. Cold water moves in to take its place. You can easily see this by heating a beaker of water with a coloured crystal in it. The path of the water is shown by the colour from the crystal – each path is called a *convection current*. Moving convection currents carry water around a central heating system.

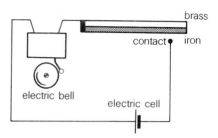

Figure 7 A bi-metal strip can be used to make a simple fire alarm.

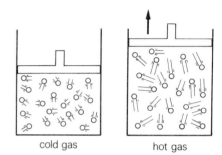

Figure 8 A gas expands when heated because its particles move faster and need more space.

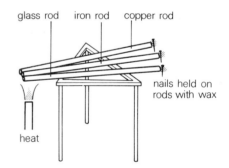

Figure 9 Heat travels along the rods and melts the wax. The pins fall off.

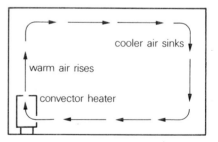

Figure 10 This moving air current is called a convection current.

Red hot objects: radiation

Heat rays If an iron nail or a piece of steel is heated enough it becomes 'red-hot'. The wire, or element, in an electric fire becomes red-hot when it is switched on. The fire gives out or *radiates* heat. You can feel this heat directly, even a few metres away. The heat travels from the red-hot wire to your skin. Heat reaches Earth from the Sun in the same way – by *heat rays*. These rays carry heat from hot objects to colder ones, even through air or empty space.

Light rays Carry on heating red-hot metal and it may even become 'white-hot'. The metal inside a light bulb (called a filament) glows white-hot and gives out light or light rays.

Heat and light rays are both important types of *radiation* which come from the Sun.

An electric fire radiates heat.

Galileo's thermometer used the rise and fall of water to give a rough idea of the air temperature.

Summary

1 When solids are heated they expand. Unless they have room to expand they push with strong forces.

2 Liquids expand about 10 times as much as solids, and gases expand about 10 times more than liquids when they are heated. Near its freezing point water actually expands as it cools and becomes ice.

3 Thermometers use the fact that liquids expand to measure temperature. Two metals can be joined to make a useful bi-metal strip, which bends when heated.

4 When things are heated their molecules move around more quickly and take up more space. This explains expansion.

5 Heat travels through some materials by conduction. Metals like copper and silver are good conductors.

6 Heat travels quickly through water and air by convection. Rising currents of air or water are called convection currents.

7 When things are heated they give off heat rays. You can sometimes see light rays from a red-hot or white-hot object.

Exercises

1 Describe two ways of showing that solids expand when they are heated.

2 Why are some bridges laid on rollers?

3 Explain why:
   a) water pipes burst if the water in them freezes
   b) fish in a pond stay alive even in very cold winters.

4 Describe how Galileo made a rough air thermometer.

5 What is a bi-metal strip? Describe how it works. List three uses for bi-metal strips.

6 Copy out and fill in the missing words:
   Heat travels through metals by _____ . It always travels from _____ parts to _____ parts. Heat can also travel in liquids and gases by _____ , and through empty space by _____ .

7 How does heat from the sun reach the Earth?

8 Which materials are the best conductors of heat?

9 Explain in your own words why:
   a) materials expand when they are heated
   b) hot air rio00
   c) convection currents occur in water
   d) a gas can expand and push a piston upwards.

Matter in outer space

The three different kinds of matter can all be seen on Earth. Most of the Earth's surface is covered in a liquid: sea-water. The rest of its surface is solid. The Earth itself is surrounded by a mixture of gases, mainly nitrogen and oxygen. Deep inside the Earth lies hot, liquid rock. But what about the rest of the Universe? You will see later that most of the matter in the Universe is probably gas.

Other parts of the solar system

Mercury is the closest planet to the Sun. It has hardly any atmosphere around it. This means there is nothing to protect it from the Sun's rays in the day or to hold the heat in at night. In the day temperatures reach 400 °C, which is hot enough to melt lead. At night the temperature on Mercury drops to nearly −200 °C. Mercury is a hostile planet.

Venus is the planet which comes closest to the Earth and it can often be seen at dawn and dusk. Unlike Mercury, Venus has a thick, heavy atmosphere mainly of carbon dioxide gas. This traps the Sun's heat like a greenhouse. The temperature on the surface is nearly 500 °C. Many of the solids you see on Earth would be liquids on Venus. And liquids, like water, would be boiled away.

Mars is mostly like a huge desert. At the top and bottom of Mars two white spots can be seen. These are 'polar caps' – they are ice caps made of frozen water and frozen carbon dioxide (often called dry ice). Mars has a very thin

Mars, showing one of its polar caps

atmosphere, with no oxygen. It has no plants or running water. People could only live on Mars in specially made 'space stations'.

Jupiter has more matter in it than all the rest of the planets put together. Most of Jupiter is hydrogen. On Earth hydrogen is a light gas. But deep in the centre of Jupiter hydrogen is crushed by this huge planet into a solid metal. The atmosphere of Jupiter is mainly made up of hydrogen and helium gas. This surrounds Jupiter in thick clouds and foggy gas so that its surface can never be clearly seen. But one thing is clear – the Great Red Spot. This spot is probably a giant hurricane which never stops blowing.

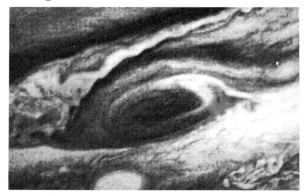

The Great Red Spot on Jupiter is probably a giant hurricane which never stops blowing.

Saturn has a mass 100 times greater than Earth's. Yet most of Saturn is gas. This planet has millions of rocks, boulders, and tiny dust particles in orbit around it. From Earth these are seen as bright rings.

All the outer planets, from Jupiter to Pluto, are extremely cold. They are far too cold to have liquid water on their surface. Pluto, often called the 'icy outpost of the solar system', is completely covered in a very cold coat of frozen material.

The Sun and other stars

The star of our solar system, the Sun, has a mass 300 000 times bigger than Earth's. Most of this mass is gas, mainly hydrogen. Nothing could remain solid on the Sun's surface. The temperature there is 6000 °C, except at the dark spots where it drops to 4000 °C. At the centre of the Sun the temperature is probably about 16 million °C.

Other stars The Sun is one of the smaller stars amongst millions in the Universe. Some are larger than others, some are hotter than others. But *all* of them are made from clouds of gas. Astronomers believe that most of this gas is *hydrogen*, which seems to be the most common substance in the Universe.

Dying stars Even stars have a lifetime. The Sun is middle-aged and only has about 5 billion years to live. When a star the size of the Sun dies it *collapses* until it is about the same size as the Earth. It becomes a very heavy heap of ash and is called a 'white dwarf star'. A cupful of matter from a white dwarf would have a mass of one hundred tonnes (1 tonne = 1000 kg). Later you will find out how some white dwarves can turn into black holes!

Visitors from outer space

Every day the Earth collides with millions of pebbles, chunks of iron, and clouds of space dust. Luckily the atmosphere shields you from this steady bombardment. Any chunk of iron is boiled away to smoke by the heat of the friction as it flies through the atmosphere.

Even so millions of particles of dust and grit get through. Every day they add another 1000 tonnes to the Earth's mass. Compared to the Earth's mass of 6000 million million million tonnes this makes no difference. But every so often a huge chunk of rock or iron reaches the Earth's surface. These chunks are called *meteorites*. About twenty-five are found each year. One was found in Arizona, USA with a mass of almost 250 kg – about the same mass as three medium-sized men. Meteorites leave large holes or craters where they hit the Earth. One crater in the Arizona desert is over 200 metres deep and 1¼ km across.

Aerial view of a meteor crater in Arizona, USA

Craters on the moon Craters on the Earth are very unusual. But there are hundreds on the surface of both Mercury and the moon, which have no atmosphere to protect them. When Galileo first saw the dark patches on the moon he thought they were seas. But really they are craters filled with dark lava from inside the moon. The moon's seas are solid!

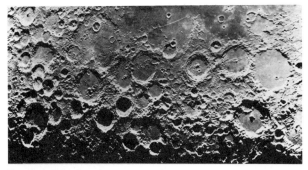
The moon's surface has many craters.

Is space empty?

Space is a dusty place. Particles of dust and atoms of gas float around between stars. This collection of space dust and gas is called *interstellar matter*. Sometimes this matter collects together in clouds which often hide the light from the stars. These dark clouds of dust and gas are called *nebulae*. They contain the simple molecules needed to make up the large, complicated molecules inside all the living things on Earth. Perhaps the Earth, and life itself, was formed when a thick cloud of dust and gas 'condensed' to form our planet.

The Horsehead nebula is a dense cloud of gas and dust.

Topic 1 Exercises

More questions on matter

1 Explain why:
   a) at the top of a high mountain it takes longer to boil an egg than it does at sea-level
   b) salt is often spread on the roads in winter
   c) you often see condensation on a cold bath tap
   d) concrete roads are laid in sections, with gaps between them
   e) a steel needle can be made to float on water.

2 The world long-jump record was broken in the 1972 Olympics in Mexico. How was the long-jumper helped by the conditions there?

3 This diagram shows an ice cube floating in a beaker completely full of water. What happens to the water level when the ice melts? Explain your answer.

Things to do

1 **An oil-propelled fish**

Draw a fish on a piece of paper. Cut it out with a channel from its tail to its middle. Then lay the fish flat on water in a wash basin or sink. The upper side must be quite dry. Now put a tiny drop of oil into the hole shown. The fish begins to move forwards.

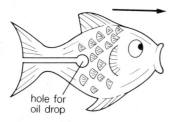
hole for oil drop

Why? Oil always spreads over the surface of water, as thinly as possible. With this fish it can only spread *backwards* along the channel. As the oil goes backwards the fish is pushed forwards. (You will see in Topic 2 that a rocket works in the same sort of way.)

2 **Boiling water in a paper cup**

You need: a paper cup, water, candle. Hold a paper cup, with a *little* water in it, over a candle flame. After a few minutes the water boils.

The water takes heat away from the paper, which never reaches a high enough temperature to catch alight.

3 **Cutting ice**

You need: an old table knife, two ice cubes, an old table. On one ice cube press down hard, keeping the knife still. You won't cut much ice. But on the other press down and move the knife backwards and forwards like a saw. Keep sawing and eventually you will cut the cube in half.

Why? Just pressing on the ice is not enough to melt it. But rubbing to and fro makes enough heat to melt the ice. As the ice melts under the knife it cuts through. An ice skate works in the same way. As the skate rubs against the ice it melts. A skater glides along on a thin layer of water.

4 **A trick with salt**

You need: an ice cube, jar of water, short piece of thin string, and some salt.

Ask somebody if they can raise the ice cube without touching it? It can be done . . . First wet the end of the string. Then lay it on the ice cube and sprinkle salt on it. This makes some of the ice melt. But heat is needed for melting (latent heat). This heat is taken from the water on the string. The water on the string freezes solid onto the ice cube, which can then be lifted.

5 **Looking for rising heat currents**

You need: a piece of card, scissors, cotton or light string. Draw a circle on the card and cut it out. Then cut the circle into a long spiral. Connect a piece of cotton to the centre of the circle, which is one end of the spiral.

If you hold the cardboard spiral above a hot radiator it starts to spin. Try holding it over: a table lamp, a hair drier, a candle flame, a pan of hot water. It should spin. Why? Rising currents of hot air push against the spiral. They are called 'convection currents'.

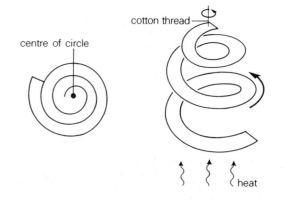
cotton thread
centre of circle
heat

TOPIC 2
Energy

A bushfire has a lot of heat, light, and sound energy

2.1 Introducing energy

You know when you haven't got any. Some people have so much it makes you tired to watch them. Where does energy come from? How can you get it when you need it?

What is energy?

People often say that somebody is 'full of energy'. The children in Figure 1 need energy for running, jumping, and climbing. Where does this energy come from? The energy of any human being comes from food, whether it is cornflakes, sweets, or an apple. But not only people have energy. So do many other things: the wind, machines, moving water, gas, electricity, and the Sun.

Try asking people what energy *is*. They may say: 'coal, oil, and gas', 'the go of things', 'what fuels have', or 'what gets things done'. Energy is needed to do jobs. A person or a machine with a lot of energy can move quickly, climb a hill, lift a heavy load, or mow a lawn. This is the meaning of *energy* used in physics: *the ability to get things done or to make things go*.

Where does energy come from?

People need food and machines need fuel to make them go. Food and fuel provide energy. So where do food and fuel come from?

On Earth almost all energy comes from the Sun. Sunlight is needed to grow the food that you eat. The fuels people use, like coal, oil, petrol, and gas were formed millions of years ago from the remains of plants and animals. Foods and fuels store energy from the Sun.

Most of the electricity that people use to provide energy in their homes is made by burning coal or oil in power stations. Figure 2 shows where Britain's energy came from in 1982. As you can see, over a third of it came from burning oil (39%).

All the different types of energy are described in the next section. Can you think of any useful energy which did not come originally from the Sun?

Different kinds of energy

Heat and light energy Two kinds of energy which come directly from the Sun every day are *heat energy* and *light energy*. Heat and light energy help food to grow, and helped to form fuels like coal and oil millions of years ago.

Chemical energy The energy stored in food and fuels is called *chemical energy*. Petrol, paraffin, wood, coal, and gas all contain chemical energy (Figure 3).

Sound energy Whenever you bang a drum, speak, or play a guitar you are making another type of energy: *sound energy*. Most rock groups make a lot of sound energy.

Figure 1 Humans get energy from food.

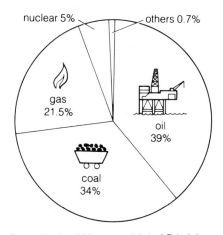

Figure 2 In 1982 over a third of Britain's energy came from oil.

Figure 3 Coal is one place where chemical energy is stored.

Nuclear energy Another kind of energy, only discovered about 60 years ago, is the energy released by atoms. This energy is released when an atom bomb explodes, as shown in Figure 4. The energy is called *nuclear energy*. In fact nuclear energy from a bomb is mainly changed into heat energy, light energy (the flash), and sound energy (the bang).

Electrical energy One of the most useful kinds of energy is electricity or *electrical energy*. It can be carried around from one place to another by wires and used where it is needed – to light a lamp, heat a kettle, or drive an electric train. Some machines which work with electrical energy are shown in Figure 5. Electricity may be the most useful form of energy, but it is very expensive.

Figure 4 A nuclear explosion releases a lot of energy.

Figure 5 These machines work with electrical energy.

Moving energy An arrow flying through the air, a moving cricket ball, a falling brick, or a person running all have energy because they are *moving* (Figure 6).

Moving energy is often called *kinetic energy* after the Greek word for movement: 'kinesis'. The faster an object is moving the more kinetic energy it has.

Stored energy Have you ever tried stretching a catapult or lifting a heavy weight? You use energy from your body to do each of these jobs. Where does the energy from your body go? It is *stored* in the catapult elastic or the raised weight. This stored energy is waiting to be used. Release the catapult and the stone flies through the air. Let go of the weight and it falls to the ground. *The stored energy changes to moving energy*.

Stored energy is often called *potential energy*, meaning 'waiting to be used'. Chemical energy can be thought of as a type of potential energy. How could the potential or stored energy in Figure 7 be used?

Moving energy and stored energy are very important in physics.

Figure 6 This runner has kinetic energy.

coiled spring

Figure 7 Energy can be stored for later use.

A dam stores water and potential energy.

Changing one kind of energy into another

The forms of energy The different kinds of energy are: heat, light, chemical, sound, nuclear, electrical, kinetic, and potential energy. These are called *the forms of energy*. All the different types of energy can be changed into another form.

'Using' energy Whenever things use energy they are really changing it from one form into another. Figure 8 shows some examples.

There are hundreds more examples of energy changes in everyday life. A television, a match, light bulb, microphone, car engine all change energy of one kind into another.

The law of conservation of energy Energy is never used up. This sounds strange, but it is true. *Whenever energy changes from one form to another the total amount of energy stays the same*. This is called the law of conservation of energy.

Energy: what's the problem?

Waste If energy is never used up why are people burning more and more fuel to supply the world with energy? Why do people talk about an 'energy problem' or an 'energy crisis'? The answer is that *most* of our energy ends up as useless *heat energy*. Heat energy is constantly escaping into the air or atmosphere around you. It escapes from houses, factories, offices, car engines, and even from people.

This heat energy in the air around you is almost useless. It cannot be used to do jobs like electrical energy can, for example. All it does is warm up the air around you very, very slightly but not enough for you to notice the difference. This heat energy is *wasted* energy.

The energy crisis People use more and more fuel to heat homes and offices, to drive cars and lorries, to run factories, and to make electricity. The fuel used – coal, oil, gas, and petrol – is becoming more and more expensive as more and more is being used up. One day all the coal, oil, and gas will have gone, as Figure 9 shows. This problem is called the *energy crisis*.

a kettle changes electrical energy into heat energy

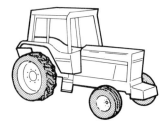
a tractor changes chemical energy in the petrol into moving energy

a wind-up toy changes potential energy stored in a spring into kinetic energy

a drum changes moving energy of a drumstick into sound energy

a car battery changes chemical energy into electrical energy

Figure 8 Energy can be changed from one form into another.

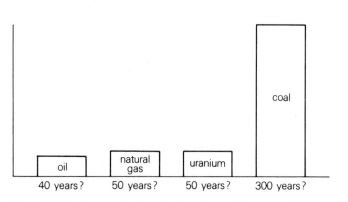

oil	natural gas	uranium	coal
40 years?	50 years?	50 years?	300 years?

Figure 9 How long will fuels last, if used up at the present rate?

New sources of energy Suppose there were no fuel left. Where would our energy come from then? Some scientists are trying to find ways of making electrical energy from the moving energy of the tides, wind, and waves. The problem is how to use these sources economically.

This wind turbine, or modern windmill, supplies enough electrical energy to the National Grid to run 200 one-bar electric fires.

Summary

1 Energy is the ability to get jobs done or to make things go.

2 Most of our energy comes from the food we eat, or the fuels we use like coal, oil, and gas. All our energy comes originally from the heat and light of the Sun.

3 There are many different forms of energy: heat, light, chemical, sound, electrical, nuclear, kinetic, and potential energy.

4 We use energy by changing it from one form into another.

5 A lot of energy ends up in the form of heat energy which escapes into the air. Fuels like coal and oil are gradually being used up. These are the reasons for the energy crisis.

Exercises

1 Copy out and fill in the missing words:
Energy is the ability to do _____ . People get their energy from _____ ; engines need _____ for their energy. The fuel used most in Britain is _____ . Fuels like coal and wood contain _____ energy which changes to _____ energy when they are burnt. The most useful form of energy is _____ energy. All moving objects have moving or _____ energy. A stretched spring or a raised weight has stored or _____ energy.

2 These three drawings show a brick being dropped. What kind of energy does the brick have in drawings **A** and **B**? What happens to this energy when it hits the floor at **C**?

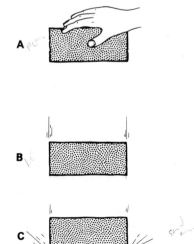

3 Fill in the right form of energy in each of these blanks:
a) A TV set changes _____ energy to _____ and _____ energy.
b) A light bulb changes _____ energy to _____ and _____ energy.
c) An electric kettle changes _____ energy to _____ energy.
d) An electric motor changes _____ energy to _____ energy.
e) An atomic bomb changes _____ energy to _____ , _____ , and _____ energy.
f) A candle changes _____ energy to _____ energy.

4 This diagram shows someone about to fire a catapult. What kind of energy does the stretched elastic have? If the person lets go describe, in your own words, how this energy changes.

5 Write down the one that is not a form of energy:
a) heat b) light c) weight d) sound

6 Explain how energy is wasted. What is meant by the 'energy crisis'?

2.2 Energy converters

Where does all your energy go? It is always being changed or converted from one form into another. Things which change energy from one form into another useful form are called energy converters. Whenever you change the energy in a hamburger into a sprint to catch the bus you are acting as an energy converter.

Living energy converters

Plants and animals, including human beings, all change energy from one form into another.

Plants Plants convert light and heat energy from the Sun into energy for growing. As they grow the plants store more and more chemical energy.

Animals Animals often eat plants to supply them with energy, Food from plants is one type of chemical energy. Animals need this energy to grow, to move around, and to keep warm.

Humans Humans obtain their energy by eating food: plants, animals, or both. Whenever you run, walk, ride a bicycle, or jump up and down you change some of this food energy into moving or kinetic energy, as shown in Figure 1. Exercise makes you hungry. But if you *take in* more energy from food than you *use* your body might store this energy and become fat. Once you stop growing your weight stays the same as long as the energy your body uses is roughly balanced by the chemical energy in food it takes in.

Waste Some of your energy is always wasted. When you run around your body becomes hot and loses heat energy. All energy converters make some energy which is of no use, usually unwanted heat or sound energy.

The next sections are all about *non-living energy converters*.

Cells and batteries

A *cell* changes chemical energy into electrical energy. A *battery* is made by putting two or more cells together.

The first cells In 1786 an Italian called Luigi Galvani thought that frogs could make electricity. He hung dead frogs from a copper hook and noticed that if they touched an iron railing they twitched violently. Galvani believed that the frogs held a store of 'animal electricity', which was somehow released when they touched two different metals.

Then in 1800 another Italian called Volta said that really this electricity came from the metals. He placed a metal on each side of his tongue, as Figure 2 shows, and found that this made a sharp taste if the metals touched each other. Volta realized that a cell can be made with two metals and a chemical between them. The chemical didn't have to be inside a frog! He built a pile of silver and zinc plates separated by salt water, something like Figure 3. It is called a *Voltaic pile*. This was the first collection of cells – the first battery.

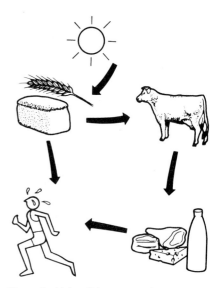

Figure 1 Living things convert energy from the Sun into other forms.

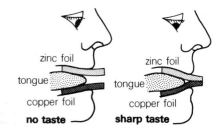

Figure 2 The tongue can form part of a simple circuit.

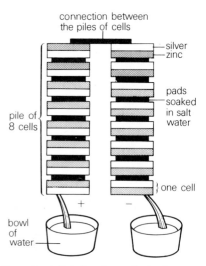

Figure 3 A Voltaic pile made the first battery.

Simple cells Figure 4 shows how a simple cell can be made with a lemon. The chemical energy from the lemon is changed into electrical energy which lights the bulb. A cell can also be made by placing the copper and zinc plates inside weak sulphuric acid in a beaker. The cell lights a small bulb. But after a few minutes the bulb goes out. Bubbles appear on the copper plate, and these stop the electric current. You can even make electricity with a jar of tap water, a 2p coin, and a 10p coin. But the current is so small that you need a very sensitive instrument to detect it.

Useful cells These simple cells are easy to make but they do not make a steady, strong current. A more useful cell is used to make up a car battery. This uses lead plates dipping into sulphuric acid. It can be 're-charged' and used over and over again.

The small cells and batteries used in radios, torches, and tape-recorders contain a paste instead of a liquid like acid. They are called *dry cells*. They are useful because they can be carried around without spilling, but they cannot be recharged.

All these cells change chemical energy into electrical energy. One new type of cell can change energy from the Sun's rays directly into electrical energy. This cell is the *Solar cell*. It is already used to supply satellites with energy. Solar cells may be used more and more in the future to provide electrical energy.

Dynamos

In fact most of the electrical energy people use at home comes not from cells but from *dynamos*. In a dynamo a coil of wire spins round between the poles of a strong magnet to generate an electric current. Moving energy is converted to electrical energy.

Electrical energy can be converted into other forms of energy. An electric fire, a light bulb, a loudspeaker, and a motor are all electrical energy converters. Electricity and its uses will be more fully described in Topics 5 and 6.

Engines

Heat energy can be used to drive certain types of engines. Any engine that changes heat energy into movement or kinetic energy is called a *heat engine*.

Steam engines The first heat engines were steam engines. If you boil a kettle full of water, and leave it boiling, the steam will push the lid of the kettle up. A man called James Watt, in 1762, showed how this steam pressure could be used to drive an engine. Heat energy is used to make steam which takes up about 1600 times as much space as the same mass of water. The steam tries to escape and pushes against a piston. By controlling the steam so that it pushes first one way and then the other, the piston is made to move backwards and forwards. If the piston is connected to a wheel with a special connecting rod, the wheel can be made to spin round and round. This is how the first railway engines were driven.

Modern heat engines Cars, rockets, and jets are the heat engines of this century. They have all changed our lives in many ways, but

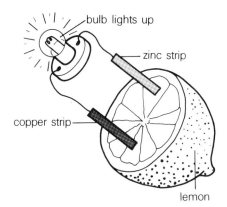

Figure 4 A simple cell can be made from a lemon.

The crossed solar panels contain solar cells which supply Skylab with electrical energy.

The steam engine is a heat engine. Pressure from the steam pushes a piston and drives the railway engine.

there are two big problems:
1. All heat engines need fuel of some sort (wood, coal, oil, or petrol). The number of heat engines is increasing all the time. The amount of fuel available is decreasing – eventually it will run out.
2. Heat engines are very wasteful. Most of their heat energy escapes into the air, instead of being changed into kinetic energy or movement. Heat engines are partly to blame for today's energy crisis.

The next two sections are all about these heat engines.

The car engine

The car engine converts chemical energy from petrol into kinetic energy. Figure 5 shows how. The engine has a container or *cylinder* with a piston inside it. The cylinder has two openings controlled by valves. As the piston moves *down* the inlet valve opens and sucks petrol and air in. Both openings are then closed by the two valves and the piston moves *up*, squeezing or compressing the petrol and air. Suddenly a spark from the sparking plug makes the petrol explode at just the right moment. The piston is pushed back *down* with a terrific force. When the piston moves *up* again it pushes out the used or burnt petrol through the second opening, the exhaust. For every explosion the piston moves up and down four times. These are called the four cycles or strokes of the petrol engine: the *intake* of petrol, *compression, explosion*, and *exhaust*.

This is a simple explanation of how a car engine changes chemical energy into moving energy. Of course the engine needs many more parts: something to feed the petrol into it (the carburettor), an electricity supply to make a spark, a system of wheels and levers to open the valves at the right time, and a way of cooling the engine down. Then a car needs some way of connecting the up-and-down movement of the piston to the wheels that drive the car. A whole book could be written on the different parts of the motor car.

Jets and rockets

Have you ever tried blowing up a balloon and then letting it go? As the air rushes out the back, the balloon is pushed forwards. In the same way, jets and rockets move forward when hot gases rush out the back of the engine. Firework rockets are pushed forward and upward in the same way.

Jets   Figure 6 shows a jet engine.

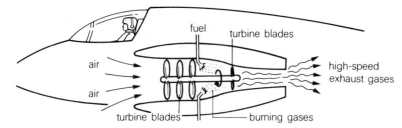

Figure 6   A jet engine is forced forward by high-speed exhaust gases.

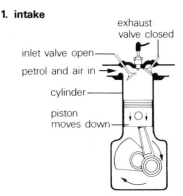

1. intake

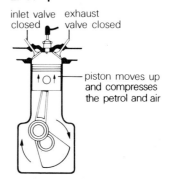

2. compression

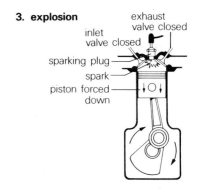

3. explosion

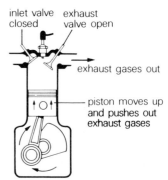

4. exhaust

Figure 5   A petrol engine is a four-stroke engine.

38

In a jet, spinning turbine blades suck air into the engine. There it is mixed with the fuel, sprayed in through a fuel pipe. The fuel burns fiercely in the air. The hot burning gases expand and are forced out of the back of the jet at very high speed. Like a balloon, the jet is forced forward as the gases rush backwards.

Rockets Every fuel needs the oxygen in the air before it will burn. But rockets travel in space where there is no air or oxygen. They must carry their own supply of *liquid oxygen*, as Figure 7 shows. The fuel (usually liquid hydrogen) mixes with the oxygen in the combustion chamber of the rocket, where it burns violently. The hot burning gases expand with the intense heat and rush out through the nozzle at the back of the rocket.

These rockets obey exactly the same laws as the ones you see on November 5th. The main difference is that they carry their own liquid oxygen and use hydrogen instead of gunpowder. Just like every heat engine they convert:

chemical energy $\longrightarrow$ heat energy $\longrightarrow$ kinetic energy
(fuel)   (burning)   (movement)

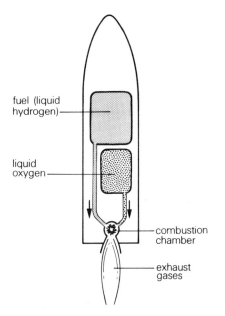
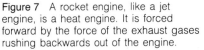

Figure 7 A rocket engine, like a jet engine, is a heat engine. It is forced forward by the force of the exhaust gases rushing backwards out of the engine.

fuel (liquid hydrogen)

liquid oxygen

combustion chamber

exhaust gases

Solar panels on the roof of this house in Milton Keynes absorb energy from the Sun

Summary

1 All plants and animals convert energy from one form to another.

2 Cells and batteries convert chemical energy into electrical energy.

3 Electrical energy is used whenever we change it into another form: with an electric fire, a motor, a light bulb, or a loudspeaker.

4 Heat engines burn a fuel to make heat energy which is then used to make things move. Heat energy changes to kinetic energy.

5 The car engine uses the chemical energy stored in petrol to drive a piston up and down.

6 Jets and rockets are pushed forwards as hot, burning gases rush out of a nozzle at the back of the engine.

Exercises

1 Make a table with the headings 'Living energy converters' and 'Non-living energy converters'. Give six examples of each.

2 Explain why plants, animals, and humans are energy converters.

3 Explain how:
a) the first battery was made
b) a cell can be made from a lemon
c) cells and batteries are used in everyday life.

4 How is most of the electrical energy that you use at home made? Write down what this energy is converted to in:
a) a light bulb
b) an electric kettle
c) a toaster
d) a spin-drier
e) an electric drill.

5 What is meant by a 'heat engine'?

6 Draw simple diagrams to show the four strokes of a petrol engine. Try to explain what happens on each stroke.

7 What is the difference between a rocket and a jet engine? How are they alike?

8 Explain with simple diagrams how a jet engine works and how a rocket works.

2.3 Saving energy

Most of the energy you use at home ends up as heat energy. This heat energy usually disappears into thin air. Because it is lost new energy must be obtained from coal, oil, and gas. These energy supplies won't last forever.

The way that heat travels

To understand how to save heat energy you need to know how heat travels. Heat travels in three different ways: by conduction, by convection currents, and by heat rays or radiation.

Conduction   Look at the saucepan in Figure 1. It is made of a metal that is a good carrier of heat. The heat energy from the cooker travels quickly through the saucepan into the water. Good carriers of heat are called good **conductors**. Most metals are good conductors of heat. The movement of heat along a material, from hot parts to colder parts, is called **conduction**.

Insulators   The handle of a saucepan is usually made of plastic. Plastic is a poor conductor of heat – it is an *insulator*. The plastic stops the heat travelling from the hot saucepan to your hand. Here is a table showing some good conductors, and some insulators:

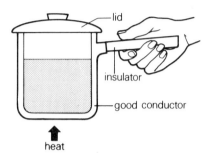

Figure 1   A metal that is a good conductor is used in a saucepan.

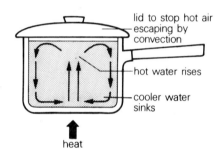

Figure 2   The water in a saucepan is heated by convection currents.

good conductors	poor conductors (insulators)	
most metals such as:	cork	air
copper	water	vacuum
silver	glass	polystyrene
steel	wood	fibreglass
aluminium	plastic	

In the next sections you will see how insulators are very useful for saving energy.

Convection   Heat is carried through liquids and gases by rising currents called **convection currents**. Convector heaters use these currents of air to warm a room. Hot air rises, cooler air sinks to take its place. The water in a saucepan is also heated by convection currents. Water circulates around the saucepan as hot water rises and cooler water sinks (Figure 2). Heat energy can be saved by using a lid – this stops convection currents from carrying hot air away from the pan.

Radiation   Heat can travel *directly* from one place to another by heat rays, or *radiation*. Radiation carries heat quickly from the Sun to the Earth through empty space and then through air. All hot objects radiate heat. But some are better radiators than others.

Look at Figure 3. Dark, dull surfaces are good radiators. Shiny surfaces are poor radiators. Shiny, highly polished teapots and kettles can save heat energy. They give away heat rays more slowly than containers with dull, black surfaces.

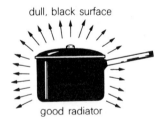

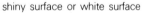

Figure 3   Some surfaces are better radiators than others.

The energy-saving house

People spend a lot of money every year heating their homes. Figure 4 shows where most of the heat goes — through the walls, the windows, or the roof into the air. They could save money, and energy, by stopping this heat from escaping.

Roof and ceiling   The heat lost through the roof can be slowed down by laying a thick layer of fibreglass in the loft. The fibreglass is a poor conductor of heat. It is an insulator. When the warm air in the house rises the fibreglass stops the heat travelling quickly to the cold air above. The ceiling itself can also be covered with an insulator, like polystyrene.

Walls   The walls in newer houses are built with an 'air gap', as Figure 5 shows. They are called *cavity walls*. The air is a poor conductor and stops heat travelling quickly from the warm inside to the cold outside. Many people cover their inside walls with wallpaper or even polystyrene. They both help to hold in the heat.

Windows   In Russia and Sweden most houses have two layers of glass in their windows, with a thin layer of air inbetween. These windows are *double-glazed*. Again the air slows down the escape of heat from the house. More and more houses in Britain use double-glazed windows but they are very expensive to buy. In fact thick heavy curtains or old-fashioned shutters are almost as good at stopping heat from escaping through the window at night.

Lagging   You can also save money by covering your hot-water tank and hot pipes with a thick layer of an insulator, as Figure 5 shows. This covering is called *lagging*. A good thick carpet on the ground floor is another kind of lagging.

These are all ways of saving heat and money. You cannot stop *all* the heat from escaping but you can certainly slow it down.

The clothes you wear

Black and white   People often wear dark clothes in winter and white clothes in summer. The experiment in Figure 6 shows you why. Heat travels from the flame directly to the white and black surfaces. The pins behind the black surface soon fall off as the wax melts. Why? The white surface *reflects* the heat rays from the flame, while the dull black surface takes them in or *absorbs* them. White and shiny surfaces are good *reflectors* of heat and light rays. Black surfaces are good *absorbers*. In summer a white shirt reflects the Sun's rays and you stay cooler. A black shirt absorbs the Sun's rays and warms you up.

Can you explain why:

    houses in hot countries have white walls?

    large petrol tanks in the desert are shiny?

    fire-fighting suits have a shiny coat?

    some people put silver foil behind their central-heating radiators?

Warm clothes   During the winter people wear thick woolly clothes or string vests and put more blankets on the bed. Birds fluff up their

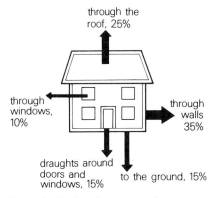

Figure 4   Heat easily escapes from a poorly insulated house.

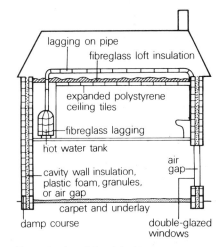

Figure 5   A well-insulated house saves heat and energy.

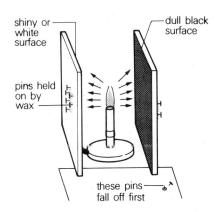

Figure 6   The dull black surface absorbs heat more quickly.

feathers, sheep and dogs grow a thick coat of wool or hair. In all these examples the covering holds or traps air inside it. Air is a bad conductor of heat, a good insulator. When trapped in your clothing it stops heat escaping from your warm body to the cold air.

A hat or a thick mop of hair can help keep you warm. People lose a lot of heat through their heads. The head heats the air above it which rises as a convection current, cold air takes its place, more heat is lost and so on.

Thermos flasks

Figure 7 shows one way of controlling conduction, convection, *and* radiation. You can use a thermos flask to keep coffee, tea, or soup hot, or to keep ice-cream and cold drinks cold. It stops most of the heat travelling out of the flask, or into it, in each of the three ways:
1. The stopper and lid stop convection currents from rising above the hot liquid and carrying heat into the air.
2. The vacuum between two glass walls stops heat escaping or entering the liquid by conduction. A perfect vacuum is just empty space – it cannot conduct heat.
3. The glass walls have a bright silvery coating on both sides to stop radiation. This silvery coat reflects back any heat rays which try to escape from a hot liquid. With a cold liquid heat rays from outside cannot get in – they are reflected back.

Friction and lubrication

Friction   Figure 8 shows some rough and smooth surfaces. The force between two surfaces is called *friction*. Friction can be useful. It helps people to hold on to things, to walk, and to turn corners. Accidents happen when there is not enough friction.

But friction can often be a nuisance. Try rubbing your hand up and down on your sleeve. Your hand begins to feel warm. The energy of movement (kinetic energy) is changing to heat energy. Whenever two rough surfaces rub together they become hot. If the moving parts of a machine rub against each other they make a lot of unwanted heat energy. Car and motorbike engines get very hot partly because of friction, and need special cooling systems to cool them down. This heat energy escapes into the air and is wasted.

Lubrication   Friction is made less, or reduced, with grease or oil. Oil helps the moving parts of bicycles, cars, lorries, and all machines to slide over each other. Anything used to reduce friction is called a *lubricant*. Using lubricants can stop machines from becoming so hot. This saves a lot of energy and increases the life of the machine.

Helping the energy crisis

How you can help   There are many ways in which *you* can save energy. Here are a few easy ones:

Switch off the lights in a room if no-one is using it.

When you make a cup of tea put just enough water in the kettle. Don't fill it up.

Save your empty bottles and either return them to the shop or

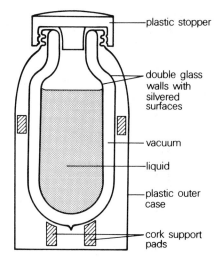

Figure 7   A thermos flask stops heat travelling by conduction, convection, and radiation.

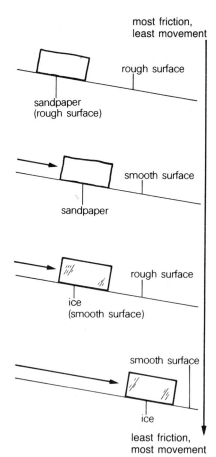

Figure 8   Friction is the force between surfaces.

take them to a 'bottle bank'. Using bottles again and again saves energy and the materials needed to make more glass.

Wear a jumper indoors and turn the heating down so that the air is cooler. You will certainly feel wider awake.

Cut the lawn with a hand-mower instead of an electric or petrol-driven one. You save energy and might even lose weight. (It saves you going jogging.)

Keep doors shut in winter. It is pointless to insulate your house and then let heat escape through the door.

Don't over-eat. Food is energy.

Save waste paper and recycle it to be used again.

Encourage people to drive smaller cars or ride bicycles.

New sources of energy But even if people save energy in all these ways we will still need new sources by the year 2000 if we want to live as we do at the moment. The section at the end of this Topic tells you about possible new sources of energy.

Bottles from a bottle bank are collected and recycled.

A well-insulated ceiling will reduce heat loss through the roof of a house.

Summary

1 Heat travels from hot places to cold places in three ways: by conduction, by convection, and by radiation.

2 Poor conductors of heat are called insulators. Insulators can be used in the walls, windows, and lofts of houses to save heat energy.

3 Many clothes are good insulators – they can save heat energy. White clothes are good reflectors, black clothes are good absorbers.

4 Thermos flasks help to stop heat travelling in each of the three different ways.

5 Some energy can be saved by oiling or lubricating machines and engines.

6 You can help the energy crisis in many ways. But new sources of energy are still needed.

Exercises

1 Copy out and fill in the blanks:
   Heat travels by _____ , _____ , and _____ .
   Most metals are good _____ . Three examples of insulators are _____ , _____ , and _____ . The Sun's rays reach the earth by _____ . Dull black surfaces are _____ radiators, but polished surfaces are _____ radiators.

2 Explain three different ways of saving heat energy in a house. How does most of the heat energy escape from a house?

3 Why do people often wear dark clothes in winter and white clothes in summer?

4 Explain why:
   a) sheep grow a thick coat in winter
   b) a string vest can keep you warm
   c) cold birds fluff their feathers up.

5 Draw a simple diagram of a thermos flask and label: the vacuum, the stopper, the silvered walls. How does each part stop the heat from escaping?

6 Describe three different ways in which you can save energy.

7 Copy the table below and complete it to show at least four good conductors, four bad conductors, and a use for each one:

good conductor	use	bad conductor	use

8 Design and sketch a 'perfect' energy-saving house. Label each part that saves energy.

2.4 Measuring energy

What have fish and chips, petrol, and electricity got in common? Why are hot things 'hot' and cold things 'cold'? Why is water 'greedy'?

This unit tries to answer questions about energy, heat, and temperature.

Joules, kilojoules, and megajoules

Joules   All the forms of energy are measured in *joules*. One joule is a very small amount of energy. It is written as 1 J. If a match is allowed to burn completely it changes about 2000 joules of chemical energy into heat energy; an ordinary light bulb changes about 100 joules of electrical energy into light energy every second.

Kilojoules   Because the joule is small, energy is often measured in *kilojoules*.

> 1 kilojoule = 1 kJ = 1000 joules

Megajoules   For even larger amounts *megajoules* are used:

> 1 megajoule = 1 MJ = 1 000 000 J

A small electric fire changes more than three megajoules of electrical energy into heat energy in an hour.

Measuring energy at home

The energy you use at home (gas, electricity, or coal) can be measured in joules:

> Electricity is sold in *units*. One unit is about 3600 kJ.

> Gas is sold in *therms*. One therm is 108 000 kJ of energy.

Gas and electricity bills are measured in therms and units because one joule is so small. Fuel bills would run into millions and millions of joules if the joule were used as the unit. One tonne (1 000 kg) of coal contains over 27 000 000 000 joules of chemical energy!

Measuring the energy that you use

Energy in food   Every time you eat you take in energy. A plate of fish and chips contains about 3000 kJ or 3 megajoules of energy. Figure 1 shows the energy in some of the other things that people eat.

Energy changes   Energy is needed for growing, moving around, working, and keeping warm. Your body changes the energy from food into other forms: kinetic energy, heat energy, and sometimes sound energy. The bigger and heavier you are the more energy you

fish and chips: about 3000 kJ

one pint of beer: 2000 kJ

one pint of milk: 1600 kJ

bowl of cornflakes with sugar: 660 kJ

one slice of bread and butter: 500 kJ

one boiled egg: 380 kJ

one apple: 170 kJ

one carrot: 85 kJ

one peanut: 25 kJ

one cherry: 10 kJ

Figure 1   Different foods give you different amounts of energy.

activity	energy used
sleeping	4 kJ
watching TV	6 kJ
walking	14 kJ
running	25 kJ
running and jumping	30 kJ
swimming	32 kJ

Figure 2   Energy used by a 16-year-old boy in one minute

need. You need more energy when you are active than when you sit still, or sleep. Figure 2 shows the energy that an average 16-year-old boy uses in one minute, doing different things.

Are you getting enough? The energy you take in every day should roughly balance the energy changed into other forms. Suppose you eat a bowl of cornflakes, a plate of fish and chips, a pint of milk, four slices of bread and butter, two eggs, and an apple in one day. Your energy *intake* is about 8000 kJ. If you are a coal miner that's not enough. A coal miner needs at least 15000 kJ of energy in a day.

Figure 3 shows the amount of energy that different people need in one day, doing different jobs. Are you getting enough energy?

person	daily energy needed
coal miner	15 000 kJ (or more)
15-year-old boy	12 000 kJ
teacher	11 000 kJ
pregnant woman	10 000 kJ
15-year-old girl	9 500 kJ
1-year-old baby	4 000 kJ

Figure 3 Different people have different energy needs.

Heat energy and temperature

Joule's experiment People once thought that heat was some kind of substance that moved or flowed from one object to another. They called it *caloric*. If a hot object touched a cold one the caloric flowed from hot to cold. In 1798 an American called Count Rumford was boring a hole in a metal cannon when he noticed that the cannon became hot. If heat were a substance, where was it coming from?

Much later J. P. Joule of Manchester did a famous experiment. As two weights fell they turned a paddle inside a tank full of water. The moving paddle actually warmed the water up. Again, where did the heat come from?

Joule explained that heat is really a form of energy, not a kind of moving substance. The energy of the falling weights and the moving paddle (and Rumford's cannon-borer) was changing to heat energy.

Temperature By measuring the temperature of something you can tell how hot or cold it is. A hot summer day might be about 30 °C, a cold winter day might be 0 °C, or even −1 °C or below. Figure 4 shows some very high and very low temperatures.

Temperature and heat are closely connected but they are *not* the same thing. Look at the two beakers full of boiling water in Figure 5. They are both the same temperature, 100 °C. But the one with the most water contains the most heat energy.

Explaining heat and temperature Heat and temperature can be explained by the Particle Theory. Look at Figure 5 again. As both beakers are heated the particles or molecules of water move around faster and faster as the temperature goes up. The higher the temperature, the faster the molecules. But the beaker with more water needs more heat energy – it has many more molecules to be speeded up. The beaker with less water reaches 100 °C first, because it needs less heat energy. It has fewer molecules to move around.

Measuring heat energy

Heating water You often heat water at home: when boiling a kettle,

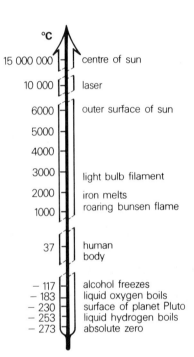

Figure 4 shows what happens at some very high and some very low temperatures.

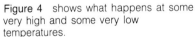

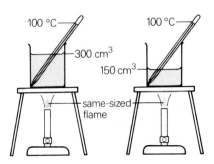

Figure 5 The temperatures are the same but the amount of heat energy is different.

45

cooking potatoes, or heating the water for a bath. Figure 6 shows how you can *measure* the heat energy supplied to water with a special meter, called a *joulemeter*. The heater in the water changes electrical energy into heat energy, just like an immersion heater in a hot-water tank. If you switch on and give the water about 20 000 joules the water is warmed up by a few degrees. Suppose you gave the water 40 000 joules of heat energy. Its temperature would then rise about twice as much.

More water, more heat energy Boiling a kettle *full* of water needs much more heat energy than if it is half-full. You can save valuable energy by not over-filling the kettle.

Heat energy and temperature rise

Other liquids Using the apparatus in Figure 6 you could heat different materials to see how much their temperature goes up. If you put 1 kg of paraffin in the beaker and give it 20 000 J of heat energy, its temperature will go up about twice as much as 1 kg of water will. You can do the same experiment with meths. The temperature of meths will also go up further than water.

Heating aluminium An immersion heater can be used to heat a block of aluminium with a mass of 1 kg. If it is supplied with 20 000 joules of energy its temperature goes up about 5 times as much as water does. It seems that water has a large appetite for heat – it is a very 'greedy' material. Different materials have different appetites for heat. Aluminium takes one-fifth as much heat as water does for the same temperature rise. It has a smaller *capacity* for heat.

Heat capacity Water can take in and store more heat than most materials – it has a large *heat capacity*. This is why hot water bottles are so good at holding or storing heat for a long time. Water is also a good liquid for central heating systems. It takes a lot of heat from the boiler to make its temperature rise, but it can store this heat for a long time in a radiator and give out more heat when it cools down.

Living near the sea If you live near the sea you often find that the winters are not as cold as further inland. The sea water stores heat from the summer sun and holds it for most of the winter. The sea changes its temperature very slowly. This means the sea can feel very cold when you swim in it during the summer, even on a hot day. The air heats up much more quickly than the sea. But at night the sea is often warmer than the air, as Figure 7 shows.

Comparing heat capacity

Specific heat capacity Some materials have a high capacity for heat, like water; some low, like aluminium. To *compare* the heat capacity of different materials you always talk about *one kg* of material being heated through *one °C*. For example, it takes 4 200 joules of heat energy to warm 1 kg of water by 1 °C. It takes only 900 joules of heat to warm 1 kg of aluminium by 1 °C. Figure 8 shows how much heat energy different materials need to warm 1 kg by 1 °C. These amounts are called their *specific heat capacity*.

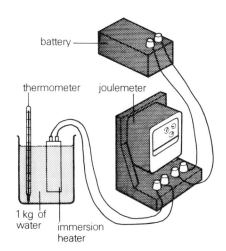

Figure 6 A joulemeter measures the amount of heat energy supplied.

Figure 7 The temperature of the sea changes very slowly.

In order to heat up by 1 °C		
1 kg copper	needs	400 J
1 kg glass	needs	670 J
1 kg concrete	needs	850 J
1 kg aluminium	needs	900 J
1 kg meths	needs	2500 J
1 kg water	needs	4200 J

Figure 8 The specific heat capacity of six different substances

Calculating heat If you know the specific heat capacity of a material you can work out how much energy it needs to heat it up. Water needs 4200 J to warm 1 kg by 1°C. How many joules will it need to heat it by 2°C? It needs twice as much: 2 × 4200 or 8400 J. To warm it by 4°C it needs 4 × 4200 or 16800 J; for 6°C it needs 25200 J, and so on (Figure 9). These are simple heat calculations.

A simple formula These calculations can be done by using a simple formula:

heat energy needed	=	mass in kg	×	specific heat capacity	×	change in temperature in °C

There are many useful formulae in physics. This is the most useful one in the study of heat or *calorimetry*. (Can you see where this name comes from?)

Summary

1 Energy is measured in joules. One joule is a very small amount of energy so kilojoules and megajoules are often used:

1 kJ = 1000 J

1 MJ = 1000000 J

2 The energy in food is often measured in kilojoules. People change this energy into other forms. These changes can also be measured.

3 Heat is one kind of energy, and so it is measured in joules. Temperature is measured in degrees Celsius (°C).

4 Water can be heated with a special heater called an immersion heater. The heat energy given to it is measured with a joulemeter.

5 Water needs a lot of heat energy to raise its temperature. It has a large capacity for heat.

6 The heat energy needed to warm 1 kg of a material by 1°C is called its specific heat capacity, measured in joules.

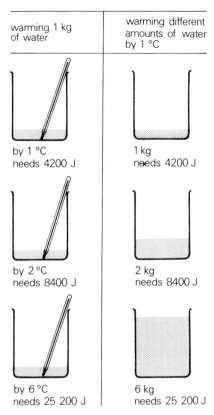

warming 1 kg of water

by 1 °C
needs 4200 J

by 2 °C
needs 8400 J

by 6 °C
needs 25 200 J

warming different amounts of water by 1 °C

1 kg
needs 4200 J

2 kg
needs 8400 J

6 kg
needs 25 200 J

Figure 9 Simple heat calculations can be made if the specific heat capacity is known.

Exercises

1 Work out the number of joules in:
a) 1 kJ b) 1 MJ c) 2.5 kJ d) 2.5 MJ
e) 1 unit of electrical energy f) 1 therm of gas
g) 1 tonne of coal.

2 Suppose you eat 20 peanuts. How much energy do these give you? How long would it take to change this energy by:

a) running? b) watching TV?
c) swimming? d) sleeping?

3 What units are used to measure:
a) temperature? b) heat?
What is the difference between temperature and heat?

4 Explain why:
a) water can be called a 'greedy' material
b) winters are often milder if you live near the sea
c) the sea at night may feel warmer than the air
d) water is a good liquid for central heating systems.

5 What does 'specific heat capacity' mean?

6 What is the specific heat capacity of: a) water?
b) aluminium? c) concrete? d) glass?
Which material is the 'greediest'?

7 Describe how scientists' ideas about heat have changed in the last 200 years. Which observations and experiments made them change their ideas?

Something for nothing?

Most of the energy people use at the moment comes from burning fuels such as coal, oil, and gas. One day these fuels will be used up. Where will our energy come from then?

All around you there are supplies of energy that can never be used up: the Sun's rays, the wind, the tides, the energy of waves on the sea. Many people believe that these are the perfect supplies for the future. They will last forever, and can be used over and over again. They are called *renewable* sources of energy. But can they really give us *something for nothing?*

Here comes the Sun

The Sun should carry on shining for at least another 5000 million years. Enough energy from the Sun falls on the Earth every *day* to provide all the energy that the human race needs for a *year*. It seems too good to be true. So where's the catch?

Collecting the Sun's energy is one of the problems. Large *solar panels* have been built in France and the United States to catch the Sun's energy. But they are very expensive to build. Smaller solar panels are already being used in the roofs of some people's houses. They can be used to warm their water. But it takes about 30 years of use for one of these to pay for itself in Sunny Britain.

Storing the Sun's energy is another problem. *Most* of our sunshine comes when we *least* need energy – in the summer. The summer Sun's heat could be stored in giant, insulated water tanks until winter. But the storage tank for a house would need to be almost as big as the house itself!

In some parts of Pakistan, solar electric pumps are being used for irrigation.

Is the answer blowing in the wind?

Two or three hundred years ago windmills were used to provide energy all over Britain. Most were used for grinding corn. But after the steam engine was invented, in 1712, windmills started to disappear. Moving energy was provided by burning coal, instead of using the energy of the wind. Now windmills are on the way back.

One good thing about the wind is that it blows *most* when we *most* need energy – in the winter. Small windmills can generate 200 watts of electrical power – enough to run a fridge and two 40 W light bulbs. Larger wind-generators can make as much as 1 000 000 watts.

But there are problems. A lot depends upon where you live. Some places, like the west coast of Britain, are windier than others. The wind does not always blow. Some way of storing 'wind energy' is needed.

Life from the ocean waves

The waves on the sea are made by the wind. Why not use 'wave energy' to help the energy crisis? Special 'ducks' called *Salter ducks* have been made which rock to and fro on each wave.

Scale model of Salter ducks

Their moving energy can be used to generate electricity. These huge ducks could be used in many of the stormy seas around Britain. Some scientists believe that if these ducks stretched for one thousand kilometres in a stormy part of the sea they could provide half of Britain's electricity. And, as with the wind, there is *most* wave energy when it is *most* needed – in the winter months.

But there are problems. The ducks are very expensive to build. And it costs a lot to carry the electricity back to land.

The rise and fall of the tides

The renewable energy looked at so far has come from the Sun: solar, wind, and wave energy. What about the tides, which come mainly from the pull of the moon's gravity? At high tide water rises. At low tide it falls again. This falling water can be used to drive turbines and generate electricity.

A tidal power station has already been built in France. It cost a lot of money to build. But it needs no fuel. In Britain, scientists are designing a tidal power station for the Bristol Channel. This could provide more than one-tenth of our electricity.

Part of the tidal power station in France

Unlike wind and wave energy, tidal energy is reliable! As long as the moon stays where it is, the sea will rise and fall every day. But there is one big problem. Interfering with the tides in the Bristol Channel could destroy many of the animals and plants that live in it and beside it. Without the regular tides these animals might die. Even the tides won't provide something for nothing.

Under our feet: geysers

Underneath every part of the Earth there is red-hot, molten rock. In some parts this molten rock overflows onto the Earth's surface when a volcano erupts. In other places water gets down to the hot rock and is sent back up again, hot and steaming. This makes a *geyser* or a *hot spring*.

Some countries have used the hot water and steam to heat nearby houses. But a country like Britain has no natural geysers or volcanic areas (though it does have hot springs or *spas* at Bath, Cheltenham, and Buxton). To use the Earth's heat *artificial geysers* have to be made. This can

be done by drilling several kilometres down into the hot rock. Cold water is sent down into the hole – it comes back up as hot water, or even steam. At the moment this is more expensive than drilling for oil. But oil will run out one day. Some scientists believe that the Earth's own store of heat will supply most of our energy in the next century. And it will never run out.

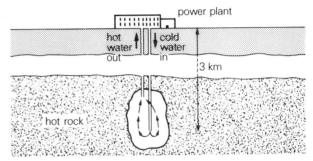

An artificial geyser uses the heat of the Earth.

'Waste not, want not': rubbish and bio-gas

Every year people in Britain throw away more than 20 million tonnes of rubbish. Burning this rubbish would supply as much heat as burning 6 million tonnes of coal. This heat could be used to run a power station. Rubbish is already used to make electricity in Germany and Holland.

The gas from rotting food, vegetables, plants, and even sewage may be smelly but it can be burnt. This 'bio-gas' may be a valuable fuel in the future. Some countries even grow special plants which are used to make bio-gas. Sugar beet and cane can produce another fuel: *alcohol*. One day this may be used, instead of petrol, to run special 'alcohol cars'. Beware of drunken drivers!

What's the answer?

All these energy supplies have a catch: the Sun does not always shine, the wind drops, wave energy and the Earth's heat are a long way from where they are needed. So which renewable energy supply is best? In a windy area, like the west coast of Scotland, wind energy could supply most of the electricity. In sunny countries, solar power stations are worth building. In places where the heat inside the Earth can be reached easily, towns and cities could have a renewable supply. All the different sources of energy could help – different sources for different places.

Topic 2 Exercises

More questions on energy

1 A stiff spring is squeezed, or compressed, by pushing a trolley against a wall:

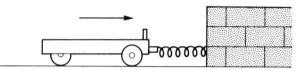

a) What sort of energy does the spring have when it is compressed?
b) What happens to the trolley when it is released?
c) After a while the trolley will slow down and stop. What has happened to the energy in the spring?

2 In the 19th century many scientists tried to build machines that carried on moving forever, once they had been started. They called these endlessly moving machines 'perpetual motion machines'. Try to explain why nobody will ever be able to make one.

3 Which one of the following (A,B,C,D, or E) shows correctly the energy changes when an electric torch, with a battery, is switched on?

A. electrical → light → heat and chemical
B. chemical → heat → electrical and light
C. chemical → electrical → heat and light
D. electrical → heat → chemical and light
E. heat → electrical → light and chemical

4 This diagram shows how electricity is made from coal, then used to light a house. Describe how energy has changed from one kind into another, starting with the coal and ending with the light bulbs in the house.

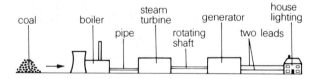

5 Life on a Mars

A well-known chocolate bar contains about 14 grams of fat and 39 grams of carbohydrate. Each gram of fat contains 38 kJ of energy; each gram of carbohydrate contains 17 kJ of energy.
a) How much energy does the chocolate bar contain?
b) Roughly how long would it take you to use this energy by
   i) sleeping?
   ii) swimming?
   iii) walking?

Things to do

1 Making a cotton reel tank

You need: a used cotton reel, a candle, three matchsticks, a knife, and a small rubber band.

Cut a thin slice of wax from the end of the candle. Bore a *small* hole in the centre of the wax. Then cut a groove in the wax.

Put the rubber band through the hole and slip a matchstick through it.

Now pull the other end of the band through the cotton reel and hold it with half a matchstick. Place another matchstick against this half by pushing it into one of the channels in the cotton reel:

The tank is ready. Wind it up by turning the matchstick in the wax groove. The rubber band gets twisted and stores potential energy. As the tank moves forward this changes to kinetic energy.

2 A balloon rocket

You can make a simple model of another type of energy converter: the rocket.
You need: a long thin balloon, the inside of a toilet roll (cardboard tube), string, and sellotape.

Stick two small loops of sellotape onto the cardboard tube. Thread a long piece of string through the middle and attach it to a door handle. Blow up the balloon and *hold* the end. Stick the balloon on the tube and let go. The balloon rushes forwards as the air inside it shoots backwards.

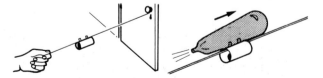

Forces

Forces go together in pairs. This frog leaping to catch a fly exerts a force on the leaf and the leaf exerts a force on the frog.

3.1 Pushes and pulls

What have a kick, a squeeze, a bend, and a twist got in common? How can an astronaut lose weight without going on a diet? The answers depend upon forces.

What is a force?

Forces are used every day in one way or another. Lifting a weight, opening a door, stretching an elastic band, squeezing a lemon, or kicking a football are all examples of forces being used. Figure 1 shows some more examples.

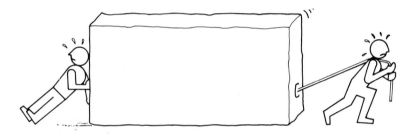

Figure 1 A push or a pull means a force is involved. All these activities involve forces.

A force is either a push or a pull. Whenever you bend, twist, stretch, or compress something you are using a push or pull: a force. Look at the drawings in Figure 2 of a block of foam being twisted and bent, squeezed and stretched. All the forces being used are either pushes or pulls.

What can forces do?

Forces can be used, or exerted, on a football, an elastic band, a lump of plasticine, a spring, or a piece of jelly. You cannot actually see these forces but you can see what they do to different objects. You see their effects. You cannot see the wind, but you can see a windmill turning, a yacht moving, or a tree bending.

Here is a list of the three things which forces can do:
- forces can change the shape or size of an object
- forces can make things move faster (e.g. a golf ball, a spring) or slower (e.g. a parachute, car brakes)
- forces can change the direction of something which is already moving (e.g. when two snooker balls collide).

The ability of a force to change the speed of an object is very useful: a car engine increases the speed of a car, a parachute slows down anything which is falling, a jet engine pushes a plane forwards. The study of speed and movement is an important part of physics called *Dynamics*, and began in the time of Aristotle (384 BC). Most of the laws on forces and movement used in physics were laid down by Sir Isaac Newton in 1679. Although more than 300 years old they are still used today.

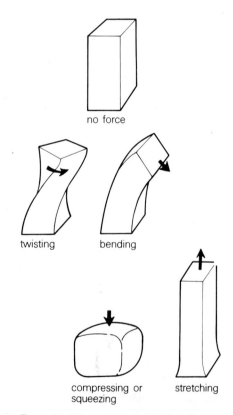

Figure 2 A force can change the shape and size of a block of foam.

Different kinds of forces

There are different kinds of forces. Examples are gravity, magnetic force, electrostatic force, friction, and tension.

Gravity   An apple falling from a tree is pulled towards the Earth by the force of gravity (Figure 3). This pulling force is called *weight*, the most common force of all, since every object on Earth has some weight. The moon has some gravity but much less than the Earth. An astronaut weighs about six times more on Earth than on the moon. In outer space, well away from the moon, the Earth, and the other planets there is no force of gravity pulling on him. This means that he has no weight – the astronaut is *weightless*.

Figure 3   Gravity is a pulling force.

Magnetic force   Another important pulling force is the force between two magnets. Two magnets can attract each other. How can you make them push each other apart? Figure 4 gives you the answer: it depends upon the positions of the ends of the magnets, the North and South poles.

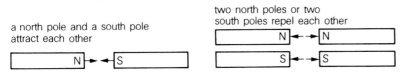

a north pole and a south pole attract each other

two north poles or two south poles repel each other

Figure 4   A magnetic force is also a push or a pull.

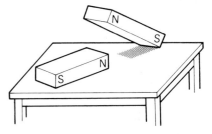

Figure 5   A magnet can be held in mid-air by the push of a magnetic force.

So magnetic forces can be either pushes or pulls. A magnet can even be held up in mid-air by a magnetic force (Figure 5).

Electrostatic force   If you rub a plastic comb on your sleeve and hold it close to a stream of water from a tap, the water will bend towards the comb (Figure 6). What force is pulling the water towards the comb? A plastic object will also pick up small pieces of paper if it is rubbed on a sleeve or a cloth. The force attracting the paper, and the water, is a third type of force: *electrostatic force*, caused by static electricity.

Have you ever tried hanging two balloons together and then rubbing each one with your sleeve? The balloons push each other apart with an electrostatic force. The same forces, from static electricity, give rise to thunder and lightning.

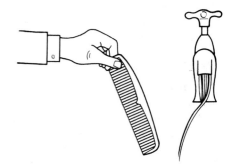

Figure 6   An electrostatic force pulls the water towards the plastic comb.

Gravity, magnetic, and electrostatic forces are all very important in physics. They all act through either *air* or *empty space*.

Friction   One important force which only acts when two objects are *touching* is the force of *friction*. Pull your hand across the top of a table – you will feel the force of friction trying to slow it down or stop it moving (Figure 7). Friction is usually a nuisance because it slows things down. In bicycles, car engines, and almost every machine the moving parts rub together and are slowed down by friction. Lubrication reduces the force of friction by putting a thin slippery layer between moving parts.

Most people think of friction as a nuisance. But without friction it would be impossible to walk, to pick things up, or for a vehicle to have brakes. Moving objects would carry on moving forever.

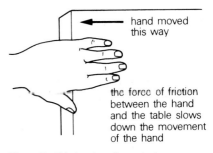

hand moved this way

the force of friction between the hand and the table slows down the movement of the hand

Figure 7   Friction is a force between objects which are touching.

Tension The forces mentioned so far are important in physics: gravity, magnetic, electrostatic, and frictional forces. One other force is important because it is useful for *measuring* the size or strength of forces. Look at Figure 8. The pulling force in the rope is called a *tension*. Whenever a string, a piece of elastic, a rope, or a spring is pulled or stretched the force used is called a tension.

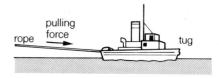

Figure 8 The pulling force in the rope is called a tension.

Measuring forces

A spring stretches when you pull it. The stronger you pull, the more the spring stretches, as long as you do not keep pulling it until it breaks. Try hanging weights on the end of a spring – as more weights are added the spring gets longer and longer. This result is known as Hooke's Law.

A simple way of measuring a force is by using it to stretch a spring. The larger the force the more the spring is stretched. A spring with a scale attached to it is called a *newton-meter*. An example is shown in Figure 9.

Forces are measured in *newtons*. Most people can pull with a force of two or three hundred newtons, while some people weigh up to 1000 newtons. This means that newton-meters are made with strong or weak springs. The meters with the strongest springs can measure the biggest forces. Some newton-meters read up to 10 newtons, some go up to 1000 newtons or more.

Weight Weight, or the force of gravity, can also be measured with a newton-meter. The more an object weighs the more it stretches the spring. An apple weighs roughly one newton, or 1 N.

The weight of an object depends upon the force of gravity pulling on it. On Earth, a mass of one kilogram weighs about ten newtons. But on the moon, where gravity is less, the kilogram-mass weighs only about 1½ newtons, as shown in Figure 10. Weight can change – mass always stays the same.

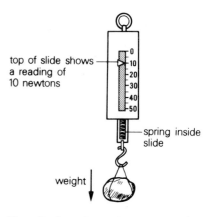

Figure 9 A newton-meter measures the force of gravity, or weight.

Forces in pairs

One more important thing about forces is that they always occur in pairs. When a girl sits on a chair, her weight presses down on the chair. The chair bends slightly but normally does not break because the *downward* force from the girl is balanced by an equal *upward* force from the chair.

Figure 11 shows some other examples.

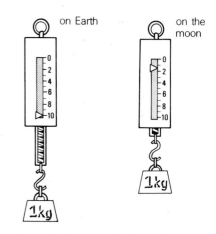

Figure 10 The weight of an object depends upon the force of gravity pulling on it.

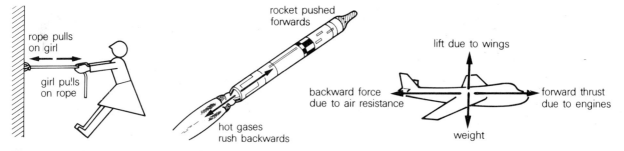

Figure 11 Forces always occur in pairs.

It was Sir Isaac Newton who first noticed that forces go together in pairs acting in opposite directions. He called the forces *action* and *reaction*.

Summary

1 A force is either a push or a pull.

2 Forces can make things move, change the size, shape, or speed of an object, or alter its direction.

3 Important forces in physics are gravity, magnetic, electrostatic, and frictional forces. Some forces, like gravity and magnetism, can act through space – some, like friction, only act when objects are touching.

4 A pulling force is called tension. It is measured by seeing how far it stretches a spring. A spring with a scale attached to it is called a newton-meter. Forces are measured in newtons.

5 The mass of an object, in kilograms, is always the same. The weight of an object, in newtons, depends upon the force of gravity pulling on it and can vary from one place to another.

6 Forces always occur in equal and opposite pairs.

This game of chance relies upon magnetic forces. The magnets on the steel base either attract or repel the magnet in the pendulum. The pendulum swings around due to the forces of attraction and repulsion.

Exercises

1 The pictures on page 52 show different forces. Look at each picture and write down whether it shows a pulling force or a pushing force.

2 Copy out and fill in the missing words:
A force can either be a _____ or a _____ .
Weight is a force due to the pull of the Earth's _____ . On Earth things weigh about _____ times more than on the moon. A plastic comb can be rubbed to make _____ electricity. When two things rub together the force between them is called _____ . Forces are measured in _____ , using a _____ – _____ . Forces go together in pairs called _____ and _____ .

3 Look at this man trying to lift the same weight on Earth and on the moon.
Try to explain, in your own words, why he finds it much easier on the moon.

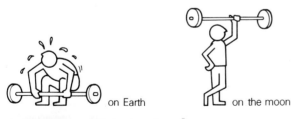

on Earth   on the moon

4 Which one of these is a force?
a) energy b) mass c) weight d) speed

5 A mass of 1 kg weighs 10 newtons on Earth. Write down the weight on Earth of:
a) 2 kg   b) 4 kg   c) 10 kg
d) 91 kg   e) 2.5 kg   f) 9.6 kg

6 This drawing shows a brick being pulled along by a rope. What is the force in the rope called? What force is slowing up the brick? How could you make the brick slide along more easily?

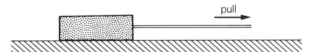

pull

7 This drawing shows two magnets held close together. In which ones (A,B,C, or D) would the magnets push each other apart, or repel? In which ones would the magnets attract?

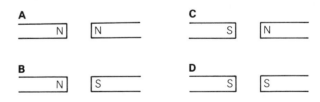

A | N | | N | C | S | | N |

B | N | | S | D | S | | S |

8 If a balloon full of air is rubbed with a dry cloth it can sometimes be made to cling to a ceiling. The reason for this is:
a) magnetic attraction
b) friction makes the balloon melt
c) balloons full of air rise
d) electrostatic attraction

9 a) An astronaut weighs 750 N on Earth. What is his mass?
b) Suppose that you weigh 500 newtons. What is your mass?

3.2 Forces, work, and power

Have you ever been asked if you know the meaning of the word 'work'? In physics you only do work when a force is moving. Doing your physics homework is not 'work'.

This unit explains three words in physics with special meanings: work, energy, and power.

Work and energy

Look at Figure 1. As the weight is falling the string turns a wheel or a pulley. The pulley is connected by a belt to a small dynamo, like a bicycle dynamo. The pulley turns and drives the dynamo which lights a small bulb. Energy is being changed from one form to another. Here is the *energy chain*:

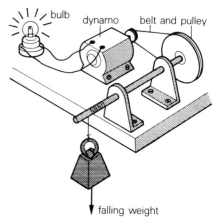

Figure 1 The falling weight starts an energy chain.

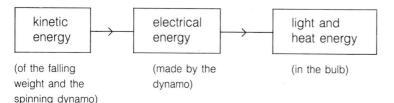

(of the falling weight and the spinning dynamo)

(made by the dynamo)

(in the bulb)

Instead of using a falling weight you could pull the string yourself and make the bulb light. You would be doing *work*. The falling weight in Figure 1 is doing work. Whenever energy is changed from one kind to another somebody or something is doing work.

Here are some examples of work:

   A girl lifting a heavy weight
   A boy climbing the stairs
   A man pushing a car that won't start
   A steam engine driving a dynamo that lights a bulb
   A crane lifting a steel girder
   A carpenter smoothing a piece of wood with sandpaper
   A girl pulling a sledge through the snow.

Some of these are shown in Figure 2.

Pushing a car involves doing work.

Figure 2 Whenever energy is changed from one form into another, work is being done.

Work has a special meaning in physics. You only do **work** when a *force is moving*. The weightlifter in Figure 3 needs a force to hold up the weight. But he is not doing any work or using up energy unless he lifts the weight up and down. The clamp holds the glued joint together and the prop holds up the wall without any work being done, since no force is moving.

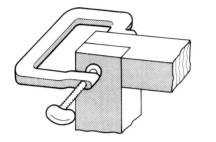 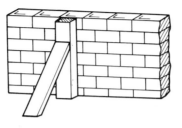

Figure 3 No force is moving, so no work is being done.

Measuring work Work can be *exactly measured* in physics. A simple formula is used:

work done	=	force used	×	distance moved
(joules)		(newtons)		(metres)

The amount of work done tells you how much energy has been 'used', or changed from one form to another. This is why energy and work are both measured in *joules*.

Here are some examples of measuring the work being done:

 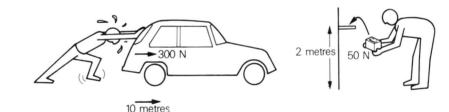

A girl pushes a trolley along for 3 metres. She uses a force of 50 newtons. The work done = 50 N × 3 m = 150 joules.

A man pushes his car for 10 metres. He uses a force of 300 newtons. He does 300 × 10 or 3000 joules of work.

A boy lifts a brick weighing 50 N onto a shelf 2 m above the ground. He does 50 × 2 or 100 joules of work. How much energy does he use? He *uses* 100 joules of energy which he *gives* to the brick.

Work and power

People often say that a machine, a man, or an animal is powerful. You hear about the power of God, power to the people, powerful cars, and nuclear power. So what does power mean? Just like the words 'work' and 'energy', 'power' has a special meaning in physics. Something is powerful if it can do work quickly. The *faster it works* the *more power* it has.

Imagine two men working as in Figure 4. They both lift bricks onto the truck. But one takes only 5 seconds to raise a brick, while the other takes 10 seconds. Which man has the most power? The man working more quickly does – he is twice as quick, and therefore twice as powerful.

Measuring power You can calculate the power of each man by using another simple formula:

$$\text{power (watts)} = \frac{\text{work done (in joules)}}{\text{time taken (in seconds)}}$$

Power tells you how many joules of work are done in one second. Power is measured in joules per second or **watts** (W for short). A machine with a power of 75 watts does 75 joules of work every second. Look again at Figure 4. If one man does 100 J of work in 10 seconds:

$$\text{His power} = \frac{\text{work done}}{\text{time taken}} = \frac{100}{10} = 10 \text{ watts} \quad (10\,W)$$

The other man does 100 J of work in 5 seconds.

$$\text{His power} = \frac{\text{work done}}{\text{time taken}} = \frac{100}{5} = 20 \text{ watts} \quad (20\,W)$$

He is twice as powerful.

Machine power Figure 5 shows a crane lifting bricks onto a lorry. Imagine that it can lift a pile of bricks weighing 4000 newtons. If it raises these by two metres,

the work done = force × distance moved
$$= 4000\,N \times 2\,m$$
$$= 8000\,J$$

Suppose this takes four seconds. The power of the crane =
$$\frac{\text{work done}}{\text{time taken}} = \frac{8000\,J}{4\,\text{seconds}} = 2000 \text{ watts}$$

The crane is much more powerful than both men put together!

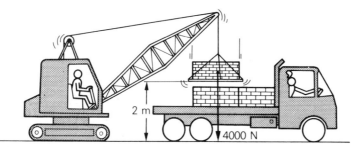

Figure 5 A crane is more powerful than several men.

Kilowatts, megawatts, and horsepower A power of one watt is very small. Larger powers are measured in kilowatts and megawatts:

1 kilowatt = 1 kW = 1000 watts
1 megawatt = 1 MW = 1 000 000 watts

Figure 4 The man who works more quickly has more power.

This machine is much more powerful than several men.

A Polaris missile is more powerful than millions of electric kettles.

The power of the crane is 2 kilowatts. Engine power is sometimes measured in *horsepower* (h.p.). One horsepower is about 750 W. Figure 6 shows the power of some of the machines and engines that people use.

Electric power

Electric lamps, drills, cookers, and kettles all change energy from one form to another. A light bulb converts electrical energy to light and heat energy. A 60 watt bulb converts 60 joules of electrical energy to light and heat every second. It has a power of 60 joules per second. All light bulbs have a certain power which is written on them in watts. It may be 40 W, 60 W, 100 W, or even 500 W.

Electric kettles are more powerful than light bulbs. Some kettles have a power of 2000 watts or 2 kilowatts. They change 2000 joules of electrical energy into heat energy every second. In two seconds they convert 4000 J of energy, in three seconds 6000 J, in four seconds 8000 J, and so on. An electric kettle uses about *twenty* times as much energy in one second as a 100 W light bulb does.

machine	power
moon rocket	100 000 000 kW
racing car	350 kW
family car	40 kW
electric kettle	2 kW
imaginary horse	750 W (1 horsepower)
man	500 W
light bulb	60 W
electric clock	3 W

Figure 6 A comparison of the power of some machines

These appliances all change electrical energy into another form. The rate at which they convert energy is called their power.

Summary

1 When you use a force to move something you do work. You can calculate how much using the formula:
   Work done = Force used × Distance moved
   Work is measured in joules.

2 The power of something tells you how fast it does work:
   Power = Work done ÷ Time taken
   Power is measured in watts.

3 A light bulb may be 40 W, 60 W, 100 W, or more; a kettle may have a power of 2000 W (2 kW). Their power tells you how quickly they convert energy.

Exercises

1 Give three examples of a person doing work. What is special about the word 'work' in physics?

2 Calculate how much work is done when:
   a) a woman lifts a tin weighing 10 newtons through a height of 3 metres.
   b) a man uses a force of 400 N to push his car a distance of 50 metres.
   c) a girl weighing 350 N jumps a height of one metre.

3 Calculate the power of:
   a) a man who does 100 joules of work in 10 seconds

b) a lorry that does 6600 J of work in 6 seconds
c) a digger that lifts 3000 N of soil by 2 metres in 5 seconds

4 How many watts in:
   a) 1 kilowatt? b) 2 kW?
   c) $3\frac{1}{2}$ kW? d) 1 megawatt?
   e) $2\frac{1}{2}$ MW? f) 1 horsepower?

5 A light bulb converts electrical energy into light and heat energy. How much energy does a 60 W bulb convert in:
   a) 1 second? b) 10 seconds? c) 1 minute?
   d) 1 hour? e) 1 day? f) 1 week?

3.3 Using forces

Forces do work. They have built the pyramids, Stonehenge, and high-rise flats. They can move cricket balls and cannon balls, cars and rockets. This unit tells you how forces can be used.

Using machines to do work

Force multipliers Figure 1 shows one simple machine being used to do work – the lever. Levers can be used to change a small force into a much larger one. A small force can be made to lift a heavy weight when a crow bar is used as a lever. All levers are *force multipliers*. They change small forces into big ones. Jacks used to lift cars are force multipliers.

The Ancient Greeks used levers. The Egyptians used an even simpler machine to build the pyramids – a ramp. A ramp or a slope can be used to raise a heavy load, using only a small force. The ramp is another force multiplier.

Changing direction Some machines help by changing the *direction* of a force. Look at the pulley in Figure 2. The man pulls *down* on the rope and the heavy bucket moves *upwards*. Simple pulleys like this are often used to raise heavy loads, on building sites for example.

The trouble with machines ... Levers, ramps, and pulleys rely on human energy. But many engines and machines get their energy from fuel. Pile drivers, cranes, steam rollers, lorries, fork-lift trucks, and diggers are a few examples. They can do work quickly, give people power, and produce large forces. But machines cannot save energy. They never give something for nothing. They need to be supplied with energy. All machines waste some of this energy because of friction between the moving parts. As people use more and more machines to do the washing up, drive them around, do their maths for them, and mow the lawn, more and more fuel is used up.

Forces making things move

Keeping going If you throw a cricket ball into the air it comes back down again. A cannonball fired from a cannon or an arrow shot from a bow travel along a curved path, and eventually stop. In everyday life nothing carries on moving forever. Until Sir Isaac Newton's time people believed that a force was *always* needed to *keep* things moving. Newton said that this was not true, in his first law:

> If something is moving it will carry on in a straight line, at a steady speed, unless a force is applied to it.

So why does a cannonball slow down when it is moving through the air? Figure 3 shows you the reason. The ball is slowed down by two

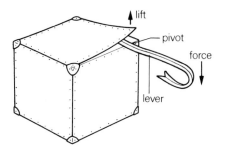

Figure 1 This lever changes a small downward force into a larger upward force.

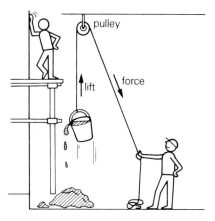

Figure 2 This pulley changes a downward force into an upward force.

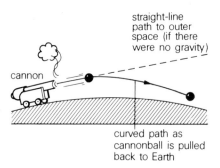

Figure 3 The force of gravity and the force of air resistance change the speed and direction of the cannonball.

forces acting on it: The Earth pulls it downwards – this is the force of gravity. The air rubs against it – this is called *air resistance*. Without these forces the cannonball would carry on in a straight line and travel into space. Rockets and spaceships will carry on moving forever in outer space where there is no air resistance and no gravity. Newton's first law might seem obvious nowadays but in his time it was a brilliant observation.

Starting and stopping   Imagine a car on a level road. To get it moving you need a force. You could push the car, or start the engine to provide you with a force. But it will not move unless it is given a push or a pull. This is another part of Newton's first law:

> If something is standing still it stays still until a force is given to it.

Once the car is moving, why doesn't it carry on forever when the engine is switched off? There must be a force somewhere. This time it is the force of friction.

Moving in a circle   Newton's first law tells us that something will carry on moving with steady speed in a *straight line* unless a force acts on it. Figure 4 shows a stone moving round and round in a circle. There must be a force pulling it round, stopping it from moving in a straight line. The force is the pull of the string. Cut the string and the stone flies off along a straight line path.

Newton's message is that whenever something is:
- speeding up,
- slowing down, or
- changing direction
there must be a force acting on it.

Can you spot the forces involved in Figure 5?

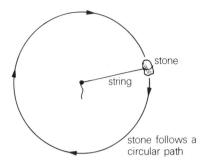

Figure 4   The force or tension in the string pulls the stone in a circular path.

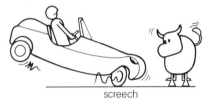

screech

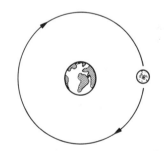

Figure 5   What forces are involved in these events?

> **Summary**
> 1   Machines are often used to do work.
> 2   Many machines are more powerful than men but they need to be supplied with energy.
> 3   Newton's first law says that a force is always needed to make things speed up, slow down, or change direction.
> 4   If there is no force acting, things will carry on moving at the same speed in a straight line, or they will just stay still.

Exercises

1   Give three examples of machines that are force multipliers. Describe how they can be used.

2   Describe a simple machine that changes the direction of a force.

3   What does Newton's first law tell you about moving objects?

4   a) What happens to a moving rocket in outer space if its engines are switched off?

b) Why does a car slow down when its engine is switched off?

c) What forces stop a moving cannonball from reaching outer space?

5   Give three examples of objects:
a) speeding up   b) slowing down
c) changing direction
What forces are involved in each example?

61

3.4 On the move

Concorde travels 300 metres a second.

You find forces wherever you find things moving – speeding up, slowing down, or going round in circles. In 1687, Sir Isaac Newton suggested three laws about the way things move. This unit tells you about movement and Newton's laws, in a simple way.

Speed and velocity

Speed   A sprinter in the Olympics runs 100 metres in about 10 seconds. Concorde travels about 900 metres in 3 seconds. A rifle bullet covers about 2700 metres in 3 seconds. The bullet has the highest *speed* because it covers the largest *distance* in the shortest *time*.

Comparing speeds   To compare the speeds of different moving things you find out how far they travel in *one* second. In one second:

The Olympic sprinter travels $\frac{100}{10}$ or 10 metres.

Concorde travels $\frac{900}{3}$ or 300 metres.

The rifle bullet travels $\frac{2700}{3}$ or 900 metres.

A simple formula can be used to work out the speed of something:

$$\text{speed} = \frac{\text{distance travelled} \quad \text{(in metres)}}{\text{time taken} \quad \text{(in seconds)}}$$

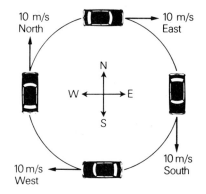

Figure 1   The speed of the car is steady but the velocity is constantly changing.

In one second the sprinter travels a distance of about 10 metres. His speed is 10 metres per second (10 m/s). Concorde's speed is 300 metres per second. A rifle bullet's speed is about 900 metres per second.

Velocity   Knowing the speed of a moving object tells you how far it travels every second. It does not tell you which way the object is moving. If you know the *velocity* of something you know *how fast* it is travelling and in *what direction*. A car can move round and round in a circle at a steady speed, as Figure 1 shows. Its direction keeps changing. Its speed is steady, but its velocity is changing all the time.

To tell somebody the velocity of a car you must say:
- its speed, e.g. 20 m/s
- its direction, e.g. north-east.

Uniform motion   A rocket in space, an aeroplane in the sky, and a car on a motorway often travel at a *steady speed* in a *straight line*. This kind of movement is called *uniform motion*. Look at Figure 2. All these objects are moving with uniform motion. The rocket is moving through space with uniform motion because there are no

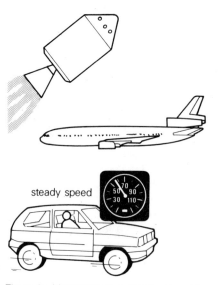

Figure 2   Movement at a steady speed in a straight line is called uniform motion.

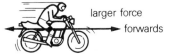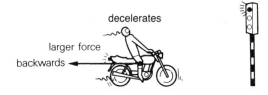

accelerates

larger force
forwards

uniform motion

equal   forces

decelerates

larger force
backwards

Figure 3 The forces are different when a bike is speeding up, cruising, or slowing down.

forces acting on it. Its engines are switched off. There is no air resistance or friction to slow it down.

The car and aeroplane are pushed back by air resistance and friction, but pushed forwards by their engines. If these two pushes balance each other, the car or aeroplane travels with uniform motion.

Speeding up and slowing down

Acceleration Look at the motorbike in Figure 3. At first it picks up speed as the force of its engine pushes it forwards. It *accelerates*. When the bike reaches a steady cruising speed the force of its engine is just enough to balance air resistance and friction. It travels with uniform motion. If the engine is switched off the forces of friction and air resistance slow the bike down. It *decelerates*.

Newton's second law If two spaceships were both pushed forward by the same force, which one would speed up or accelerate more quickly? The one with less mass would. Isaac Newton first said that *a small mass accelerates more than a large mass if both are given the same force*. This is called Newton's second law. It is summarised in Figure 4.

Inertia Racing bikes and sports cars are made with as small a mass as possible so that they can accelerate quickly. They can also slow down quickly. Steamrollers have a large mass. It takes a large force to make them speed up, and a large one to slow them down. They have a large *inertia*. Inertia means the tendency of things to keep on moving if they are already moving or to stay still if they are not moving. When you travel in a car you can see the effects of your body's inertia. If the car suddenly stops you carry on moving – this is why you need a seat belt. If the car starts off suddenly you are thrown backwards.

Moving the Earth The Earth has a very, very large mass – about 6 000 000 000 000 000 000 000 000 kilograms. It has a lot of inertia. Every time someone jumps up, or a rocket takes off, the Earth is pushed backwards as Figure 5 shows. Because it has such a big mass you do not see the Earth accelerate. But if everyone in China jumped up at the same time you *might* actually see the Earth move!

Ticker-timers

Figure 6 shows a trolley pulling a paper tape behind it as it moves along. A little machine called a *ticker-timer* puts black dots onto the tape at the rate of 50 dots every second. If the trolley moves slowly the dots are close together – as the trolley speeds up, pulling the tape behind it, the dots get further and further apart.

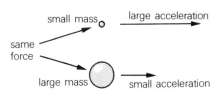

small mass   large acceleration

same
force

large mass   small acceleration

Figure 4 A summary of Newton's second law

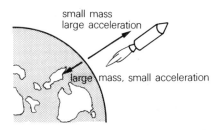

small mass
large acceleration

large mass, small acceleration

Figure 5 The effects of Newton's second law are not always obvious!

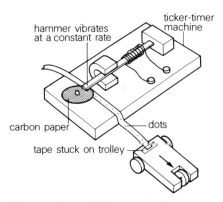

hammer vibrates
at a constant rate

ticker-timer
machine

carbon paper

dots

tape stuck on trolley

Figure 6 As the trolley moves, the ticker-timer puts dots on the paper tape.

This is how the tape looks when the trolley:

runs downhill and *accelerates*,

travels at a *steady speed*,

goes uphill and *decelerates*.

Making a chart You can make a simple chart to show how a trolley picks up speed. The chart is made by cutting the tape into strips with 5 dots on each strip, as Figure 7 shows. As the trolley accelerates each strip with 5 dots on it gets longer and longer. The chart shows how the speed of the trolley gradually increases.

Isaac Newton would have found ticker-timers very useful.

Showing motion on a graph

Speed–time graphs You can show how the speed of a car, motorbike, or any moving object changes by drawing a graph. Figure 8 shows the speed of a motorbike moving forward in a straight line. At first its speed is 0 metres per second (0 m/s). It is standing still. Then it picks up speed for 10 seconds until it reaches a speed of 20 m/s. It accelerates. For the next 40 seconds the bike travels at a steady speed of 20 m/s. It is now travelling with uniform motion. Finally the motorbike slows down quickly for 5 seconds until it is standing still again.

This kind of graph is useful for showing clearly how moving things speed up and slow down. Many people are interested in *measuring* how quickly a car can accelerate.

Measuring acceleration Figure 9 shows graphs for two different cars accelerating. The car which picks up speed more quickly has a larger acceleration. The car with the steeper graph reaches a speed of 20 m/s in 10 seconds. The other one takes 20 seconds to reach the same speed. The steeper the graph, the more the acceleration.

Acceleration is measured in metres per second per second (m/s^2 for short). You can work out the acceleration by finding how much speed the car picks up *every* second, as shown in Figure 9.

The faster car gains 2 m/s in every second. Its acceleration is 2 metres per second per second (2 m/s^2 for short). The slow car gains 1 m/s in every second – its acceleration is 1 metre per second per second (1 m/s^2).

Newton's three laws of motion

Newton's laws are about three hundred years old. But they are still used today to describe how things move. Just to remind you, here is a simple version of Newton's three laws:
- A force is always needed to start an object moving, speed it up, slow it down, change its direction, or pull it around in a circle.
- The same force will make a small mass accelerate more than a large mass. The same mass will accelerate more with a large force than with a small force.
- Forces go together in pairs, called 'action' and 'reaction'. They are always equal in strength, but act in opposite directions.

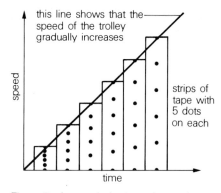

Figure 7 A speed chart can be made from the ticker-tape.

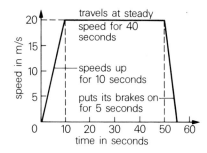

Figure 8 A 'speed–time' graph

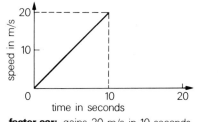

faster car: gains 20 m/s in 10 seconds or 2 m/s in every second

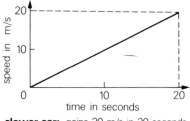

slower car: gains 20 m/s in 20 seconds or 1 m/s in every second

Figure 9 The graphs show the acceleration of each car.

Using Newton's second law   Newton's second law can be used to work out how quickly things will accelerate when a force is given to them. Figure 10 shows two trolleys, one with a mass of 10 kg, the other with a mass of 20 kg. If a push of 40 newtons is given to each one, a simple formula can be used to find out how quickly each accelerates, assuming there is no friction:

Force = 40 N = 10 kg × acceleration
Its acceleration is 4 m/s^2

force	=	mass	×	acceleration
(in newtons)		(in kg)		(in m/s^2)

The *larger* the mass the *smaller* the acceleration. Large masses have a lot of inertia.

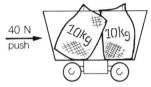

Force = 40 N = 20 kg × acceleration
Its acceleration is only 2 m/s^2

Figure 10   The same force applied to different masses gives different accelerations.

Summary

1   Speed tells you how far something travels in one second (or one hour). The velocity of something is its speed in a certain direction.

2   Large masses have a large inertia.

3   Ticker-timers are used to study the way that things slow down and speed up.

4   The changes in speed of a moving object can be clearly shown by drawing a graph or a chart.

5   Newton's second law shows the connection between force, mass, and acceleration: Force = Mass × Acceleration

Exercises

1   Work out the speed of:
   a) a woman running 400 metres in 100 seconds
   b) a bullet moving 3300 metres in 3 seconds
   c) a cheetah running 100 metres in 4 seconds
   d) a car travelling 2000 metres in 100 seconds
   e) a tortoise walking 1 metre in 5 seconds.
   List them in order with the fastest first, slowest last.

2   What does 'uniform motion' mean? Give four examples.

3   Which has the most inertia, a steam roller or a bicycle? Explain what 'inertia' means, and what it depends upon.

4   Draw simple sketches of the dots on the tape from a ticker-timer that show:
   a) acceleration   b) deceleration
   c) steady speed   d) uniform motion.

5   Sketch the shape of a speed–time graph for:
   a) a motorbike accelerating away from traffic lights, then suddenly putting its brakes on
   b) a stone being thrown up into the air, then coming down again

   c) a car picking up speed, then travelling at a steady speed for a time, then slowing down to stop.

6   Write down Newton's three laws of motion.

7   Work out the acceleration of each of these cars:
   a) a car that gains a speed of 20 m/s in 10 seconds
   b) one that gains 20 m/s in 5 seconds
   c) one that gains 30 m/s in 15 seconds
   d) one that gains 10 m/s in 2 seconds.
   Which car has the largest acceleration, which has the smallest?

8   Use the formula: force = mass × acceleration   to calculate the force needed to give:
   a) a mass of 2 kg an acceleration of 3 m/s^2
   b) a mass of 10 kg an acceleration of 2 m/s^2
   c) a mass of 750 kg an acceleration of 2 m/s^2.
   Use the formula again to calculate the acceleration of a 20 kg mass when it is pushed with a force of:
   a) 60 newtons
   b) 80 newtons
   c) 120 newtons.

3.5 Turning forces

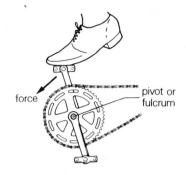

How can a heavy man and a light boy balance each other on a see-saw? Why should you not stand upstairs on a double-decker bus? Why do divers and astronauts wear heavy boots? You will find the answers in a 'moment'.

Turning effect of forces

The forces in Figure 1 are being used to *turn* something. The spanner is being used to undo a bolt. Which spanner would you use, a long or a short one? The longer spanner gives you a larger *turning effect*.

Levers A long lever can be used to lift a heavy weight as Figure 2 shows. A small force of 100 N is used to raise a large rock weighing 900 N. The long lever is a force multiplier. The lever rests on a support called a *pivot*. This is sometimes called a *fulcrum*. The fulcrum or pivot is shown in each example in Figure 1.

Magic moments With a long spanner or lever the force is a long way from the fulcrum. The further away from the fulcrum, the greater the turning effect becomes. The handle of a door is placed as far away as possible from the hinges. This makes a door easier to open. A small boy pushing near the handle can easily hold a strong man pushing near the hinges.

This turning effect is called the *moment* of force. The turning effect, or moment, of a force depends upon two things:
- the size of the force
- its distance away from the fulcrum.

You can work out the moment of force with a simple formula:

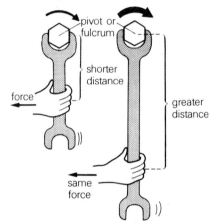

less turning effect   greater turning effect

Figure 1 Forces can be used to turn things.

moment of a force (in newton-metres)	=	force (in newtons) ×	nearest distance from the fulcrum (in metres)

Moments are measured in newton-metres (Nm for short).

Balancing

See-saws Figure 3 shows a see-saw that is balanced even though the two people on it have very different weights. The heavy man is close to the fulcrum – the girl is much further from it. They both have the same turning effect or moment.

The moment of the man is:
force × distance from fulcrum =
800 N × 1 m = 800 newton-metres.

The moment of the girl is:
200 N × 4 m = 800 newton-metres.

The two moments are balanced.

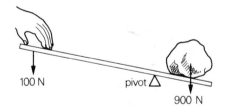

Figure 2 The small force can lift the heavy weight because the lever acts as a force multiplier.

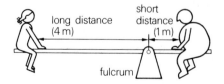

Figure 3 The see-saw is balanced because the two moments are balanced.

Balancing a plank   You can make a plank balance with two piles of bricks, as Figure 4 shows. Each pile of bricks has a turning effect. One pile of bricks tries to turn the plank round to the left. The other pile tries to pull it round to the right. The two turning effects balance each other.

Just suppose that each brick weighs 10 N. The turning effect or moment of the left-hand side is:
force × distance from fulcrum = 40 N × 2 m = 80 Nm.

On the right-hand side the moment is:
force × distance from fulcrum = 20 N × 4 m = 80 Nm.

The two moments are the same. They balance.

Here are three more examples of balanced moments:

A heavy piece of concrete balances the weight on the end of a crane.

A man carrying a big suitcase stretches his arm out to balance the turning effect of the heavy case.

Sailors on a yacht lean out over the side to balance the force of the wind.

Centre of gravity

In Figure 4 the *centre* of the plank is resting on a log. If all the bricks were taken away it would still balance. Where is the easiest place to hold a ladder when you carry it? Figure 5 gives you the answer. The ladder balances at its centre – this is called the **centre of gravity**. It is the point where all the weight of the ladder or the pull of gravity *seems* to act.

Top heavy   You can balance different objects by supporting them exactly under their centre of gravity. Many objects have their centre of gravity exactly at the middle. But some things, and even people, are called 'top heavy'. Most of their weight is near the top. They have a *high* centre of gravity. Many rugby players seem to have most of their weight in their legs. They are 'bottom heavy' and have a *low* centre of gravity.

Keeping upright

Stable or unstable   Some things fall over more easily than others. A Bunsen burner can be balanced in two positions, as Figure 6 shows. When it rests on its heavy base, it has a low centre of gravity. It does not topple over easily. The Bunsen burner is *stable*. With its base in the air the smallest touch will topple it. Its centre of gravity is much higher. The Bunsen burner is *unstable*.

Big bases   You can tell how stable something is by tilting it until it just starts to topple. Figure 7 shows a rough block of wood being tilted until it just starts to fall. If the block rests on a *wide* base it can be tilted quite a lot. On a *narrow* base the block quickly topples over.

Vases, wine glasses, milk bottles, table lamps, and stands used in laboratories all have wide, heavy bases.

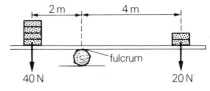
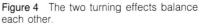

Figure 4   The two turning effects balance each other.

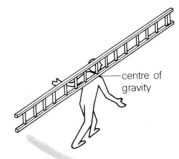

Figure 5   The ladder balances at its centre of gravity.

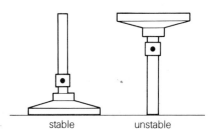

Figure 6   The Bunsen burner is balanced in both positions, but it is only stable when resting on its heavy base.

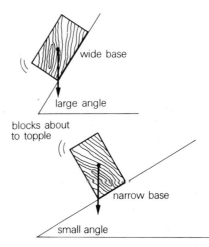

Figure 7   The block of wood can be tilted further when resting on its wide base.

Making things stable

Things can be made more stable by giving them:
- a *low* centre of gravity
- a *wide* base.

A racing car is more stable than a double-decker bus. If people stand up on the top deck of the bus it becomes less stable. Cars and buses are specially designed so that the heavy parts are kept as low as possible.

Some toys are designed with a heavy base so that they always stay upright. The toy in Figure 8 always comes back up if it is pushed over. It has a very low centre of gravity. Divers wear heavy boots to weigh them down and to help them stay upright when they walk on the sea bed.

Unstable objects are often dangerous. Figure 9 shows two examples. The centre of gravity in each case has become too high. The boat and the chest of drawers are about to topple.

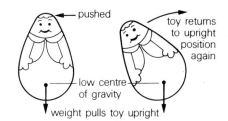

Figure 8 Because it has a low centre of gravity, the toy will always stay upright.

Figure 9 Objects become unstable when their centre of gravity is too high.

Summary

1 The turning effect of a force is called its moment. You can work it out with a formula:

moment of a force = force used × nearest distance from the fulcrum

2 A see-saw balances when the moments on each side are the same.

3 You can balance something by carrying it underneath its centre of gravity.

4 Most objects with a low centre of gravity are stable. Most objects with a high centre of gravity are unstable.

5 Objects are made more stable by giving them a low centre of gravity and a wide base. A racing car is a good example.

Exercises

1 Copy out and fill in the missing words:
The turning effect of a force is called its _____ .
It depends upon two things: the _____ of the force and its distance from the _____ .

2 Work out the turning effect of each of these forces:

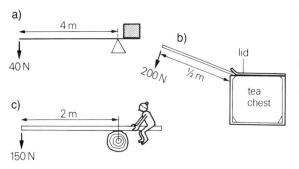

a) 4 m — 40 N

b) 200 N ½ m — lid, tea chest

c) 2 m — 150 N

Which has the largest turning effect?

3 Draw simple diagrams to show the position of the centre of gravity of:
a) a ladder
b) a 2p piece
c) a brick
d) a ruler
e) a flat cardboard square
f) a wooden plank

4 Copy out and fill in the missing words:
An object with a low _____ is more _____ than one with a high centre of gravity. An object topples easily if it has a _____ base.

5 How would you design a car to make it as stable as possible?

6 Explain why:
a) a chest of drawers is more stable if its lower drawers are filled
b) the heavy parts of a ship's cargo should be stored at the bottom
c) deep-sea divers wear heavy boots
d) people are not allowed to stand upstairs on a double-decker bus.

3.6 Under pressure

How can a man lie on a bed of nails? In what way is an elephant's foot like a stiletto heel? Why are dams thicker at the bottom than the top? The answers all depend upon pressure.

What is pressure?

Same force, different area Look at the coin being pushed into a lump of plasticine in Figure 1. Pressing on the edge of the coin makes it go in much further. The pushing force is concentrated onto a smaller area. Now look at the two men walking through snow in Figure 2. The one wearing the large, flat snowshoes does not sink into the snow. His weight is 'spread out' over a large area.

Sharp edges With your thumb you can press a drawing pin into a piece of wood. The push of your thumb is concentrated onto a very small point. A knife with a sharp edge will cut through a piece of wood. The sharp edge concentrates the force onto a very small area. The knife puts a high pressure on the wood. Blunt edges are no use for cutting. They 'spread out' the cutting force, lowering the pressure.

Spreading the weight The elephant in Figure 3 weighs about twenty times as much as the woman wearing high heels. Yet the pressure of one of her heels on the floor is as large as the pressure of the elephant's foot. Her weight is concentrated onto a much smaller area. The elephant's weight is spread over four large, flat feet. The same reasoning helps to explain why a man can lie on a bed of nails. His weight is spread out over a large number of nails. Lying on one nail would be quite painful, as Figure 4 shows.

weight is spread over many nails

weight on one nail only OUCH/!

Figure 4 Spreading the pressure reduces the pain!

Working out the pressure

The *same* force can give a high or a low pressure:

Concentrating the force on a *small* area gives a *high* pressure.

Spreading the force over a *large* area gives a *low* pressure.

same force
→ small area e.g. drawing pin ——→ high pressure
→ large area e.g. snowshoe ——→ low pressure

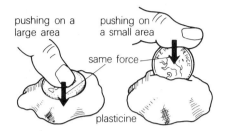

Figure 1 The coin goes in further when the pushing force is concentrated onto a small area.

Figure 2 The snowshoes spread the man's weight over a large area.

weight is concentrated at two points

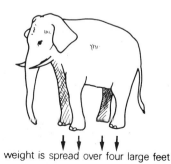

weight is spread over four large feet

Figure 3 The pressure from the woman's heel is the same as from the elephant's foot.

69

As with most things in physics, pressure can be exactly measured. You simply have to work out how much force is acting on an area of 1 m². Pressure is measured in newtons per square metre, written N/m² for short. Look at Figure 5. Each pile of blocks weighs 3600 N. The force from the left-hand pile is spread over an area of 6 m². The force acting on 1 m² is 3600 ÷ 6 = 600 N/m².

You can work out the pressure by dividing the force by the area:

$$\text{pressure} \underset{\text{(in N/m}^2)}{=} \frac{\text{force} \; \text{(in N)}}{\text{area} \; \text{(in m}^2)}$$

Sometimes pressure is measured in a unit called a pascal, Pa for short. 1 N/m² = 1 Pa.

Look at Figure 5 again. In the right-hand pile, the blocks have been rearranged. Their weight is still 3600 N but the force is being exerted over a smaller area. The pressure is
$\frac{3600 \text{ N}}{2 \text{ m}^2}$ = 1800 N/m².   The pressure is much higher.

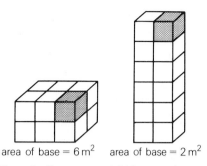

Figure 5 The same weight is spread over different areas. This makes the pressures different.

Liquid pressure

Feeling the pressure   Knives, drawing pins, and sharp heels can all be used to give high pressures. Another place where you find high pressures is at the bottom of the sea. Deep-sea divers and fish at the bottom of the ocean feel tremendous pressure. The pressure comes from the weight of the sea above them. The water presses in on the diver in every direction – from above and below, and on both sides. This is why divers need suits to protect them. At a depth of 30 metres the water pressure is about 300 000 N/m².

Getting deeper   The deeper you go the higher the pressure. Figure 6 shows a simple way of proving this. The jet of water at the bottom of the tank shoots out the furthest. For this reason, dams are built much thicker at the bottom than the top. The pressure on the bottom of a dam can be enormous, as Figure 7 shows.

Pascal's vases   People often say that 'water finds its own level'. In 1674 a Frenchman called Blaise Pascal invented a nice way of showing this. The containers shown in Figure 8 are called 'Pascal's vases'. All the containers have different shapes and areas. Yet the pressures from each one are equal because the height of liquid (the level) is the same in each vase.

Other liquids   All liquids, not just water and sea-water, have pressure inside them. Liquids that are very dense can produce very high pressures. Mercury is about 14 times more dense than water. The pressure at the bottom of a column of mercury is about 14 times higher than at the bottom of a column of water of the same height.

Using liquid pressure

Squeezing water   Have you ever tried 'squeezing' a liquid? If you fill a small plastic syringe with water and put your finger on the end,

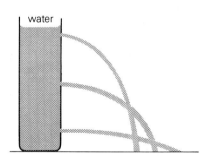

Figure 6 The higher the pressure at a hole, the faster the water spurts out.

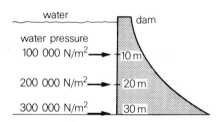

Figure 7 The deeper you go, the higher the pressure.

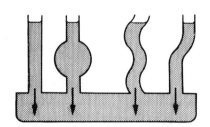

Figure 8 The pressures are the same because the water levels are the same.

you cannot push the plunger in. The water cannot be squeezed or compressed. This is shown in Figure 9.

Try holding a polythene bag or balloon, full of water. Then squeeze the top of the bag and poke holes all over it with a pin. The water squirts out in all directions. The 'squeeze' of your hand has been carried through the water. The water carries or *transmits* the pressure through it.

All liquids behave in these three ways:
- they cannot be compressed
- their pressure acts in all directions
- they transmit pressure from one part of the liquid to another.

Hydraulics Figure 10 shows how liquids can be used to make a large force from a small one. The machine is called a *hydraulic jack*. The downward force being used is concentrated on a *small* area. So it presses on the liquid with quite a large pressure. This pressure is carried through the liquid until it pushes up on the bigger piston. Now there is a high pressure pushing on a piston with a *large* area. The piston is pushed upwards with a strong force.

Machines like these are force multipliers – they can turn a small force into a large one. Car brakes, garage lifts, and some lifting jacks are all hydraulic machines that use liquid pressure.

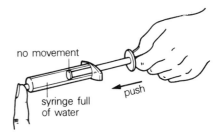
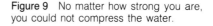

Figure 9 No matter how strong you are, you could not compress the water.

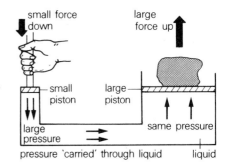

Figure 10 The hydraulic jack is a force multiplier. The small downward force at high pressure is changed to a large upward force.

Summary

1 A force pressing down on a large area makes a smaller pressure than the same force pressing on a small area.

2 pressure = $\dfrac{\text{force (in newtons)}}{\text{area (in square metres)}}$ (in N/m²)

3 The pressure inside a liquid gets higher as you go deeper. The pressure comes from the weight of the liquid directly above.

4 Pressure in liquids is used in hydraulic brakes and jacks.

Exercises

1 How can the pressure under an elephant's foot be the same as the pressure under a stiletto heel?

2 This diagram shows two blocks resting on a bench. Calculate the pressure under each one. Explain the difference.

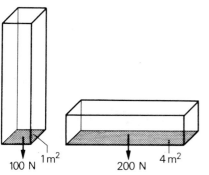

100 N 1 m²   200 N 4 m²

3 Explain why:
a) a person wearing snowshoes does not sink into snow
b) a sharp edge cuts better than a blunt one
c) a man can lie down on a bed of nails
d) a drawing pin has a sharp point.

4 Work out the pressure made by a force of:
a) 360 N over an area of 2 m²
b) 2000 N over an area of 10 m²
c) 81 N over an area of 3 m².

5 Explain why:
a) dams are built thicker at the bottom than at the top
b) deep-sea divers wear protective suits.

6 Draw a simple diagram of a hydraulic jack, and explain how it works. Give two examples of useful hydraulic machines.

3.7 Air pressure

Everyone on Earth has a weight of about 200 000 N pushing on him or her. Why can't you feel it? This unit explains the pressure that is all around you – air pressure.

An ocean of air

The atmosphere On Earth everyone lives at the bottom of a deep 'ocean' of air called the *atmosphere*. This ocean of air stretches many kilometres above the Earth, getting thinner and thinner as you rise higher and higher. The atmosphere is held close to the Earth by the pull of gravity.

Pressing down One cm³ of air only weighs about 0.001 g – so air is about 1000 times less dense than sea water. A deep-sea diver feels an enormous pressure on him from the weight of the sea above. In the same way the atmosphere above the Earth presses down on everyone. But it stretches for hundreds of kilometres above us. The pressure of this sea of air is called *atmospheric pressure*. As you rise higher above the Earth this atmospheric pressure gets smaller, as there is less and less air above pressing down.

Seeing the pressure of the atmosphere

The collapsing can Like water pressure, atmospheric pressure pushes in every direction. Why can't you *feel* this pressure? Like the tin can in Figure 1 you have air inside you as well as outside. If all the air is removed from the can using a special pump called a *vacuum pump* the can collapses. The sides are pushed in by the pressure of the atmosphere.

Magdeburg hemispheres In 1654 the Mayor of Magdeburg, in Germany, used a vacuum pump to remove the air from two small metal hemispheres pressed together to make an airtight sphere, as Figure 2 shows. Even sixteen horses could not pull the hemispheres apart, against the enormous pressure of the air.

Supporting water Figure 3 shows a simple way of 'seeing' atmospheric pressure. Hold the card in place, turn the bottle upside down, then take your fingers off the card. The atmosphere presses up on the card and supports the water in the bottle. If you used a longer and longer bottle you would find that the air pressure can hold up a column of water about 10 metres long!

Measuring atmospheric pressure

You often see weathermen on television showing areas of 'low pressure' and 'high pressure'. They try to use air pressure to forecast the weather. You can measure the pressure of the atmosphere by seeing how high a column of liquid it will support. Air pressure is strong enough to support about ten metres of water. With a much denser liquid, like mercury, the height it supports is less than one metre, as Figure 4 shows. A tube full of mercury like

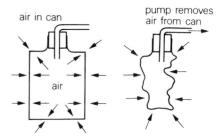

Figure 1 With no air inside it, the can collapses.

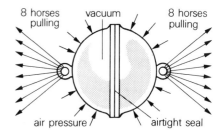

Figure 2 Sixteen horses are not strong enough to pull the sphere apart.

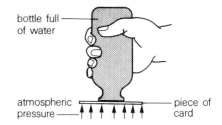

Figure 3 The air pressure is strong enough to hold the card in place.

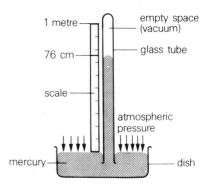

Figure 4 Normal air pressure supports 76 cm of mercury.

this can be used to measure atmospheric pressure. It is called a *mercury barometer*. Normally the atmosphere's pressure will hold up a column of mercury about 76 cm long. On some days the air pressure is low and the column may only be 75 cm long – on other days the air might hold up a column 77 cm long.

Aneroid barometers   Mercury barometers are about 15 times smaller than water barometers, but they are still big and clumsy. Figure 5 shows a smaller, portable instrument for measuring air pressure called an *aneroid barometer* (aneroid = 'without liquid'). It uses a flexible metal box with hardly any air in it, with a special spring to stop the box collapsing. If air pressure rises the box is 'squeezed in' slightly – if the pressure drops the box expands. As the box moves in or out, it pushes a pointer next to a scale.

Units   The scale in Figure 5 tells you the atmospheric pressure in N/m². Mercury barometers give you the pressure in cm of mercury. In fact a pressure of 76 cm of mercury is the same as 100 000 N/m². This is sometimes called one *bar*.

normal atmospheric pressure	= 1 bar = 76 cm of mercury = 100 000 N/m² = 100 000 Pa

Pressure can be measured in any of these units.

Climbing higher   Figure 6 shows how pressure changes as you rise higher in the 'sea' of air. These pressures are measured with an aneroid barometer by mountaineers and aircraft pilots. As you already know the boiling point of water gradually drops as you climb higher in the Earth's atmosphere. If you go down a deep coal mine, air pressure and boiling point will actually increase.

Using atmospheric pressure

Using a straw   Every time you drink through a straw you use the pressure of the air around you. As you suck on the straw you lower the air pressure inside it. The atmosphere pressing down on the drink pushes the liquid up the straw and into your mouth, as Figure 7 shows. If the top of the bottle is tightly sealed it gets more and more difficult to drink the liquid, because the air inside the bottle takes up more room and its pressure falls.

Pouring things out   Look at Figure 8. Two holes have to be made in the tin can before the milk will pour easily. The milk comes out through one hole, the atmosphere presses down through the other. Liquid pours out of a bottle more easily if you tilt it gently to allow air in, instead of tipping it right up.

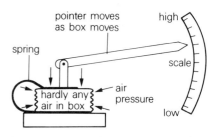

Figure 5   An aneroid barometer contains no liquid.

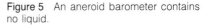

12 000 m
air pressure about 15 cm of mercury

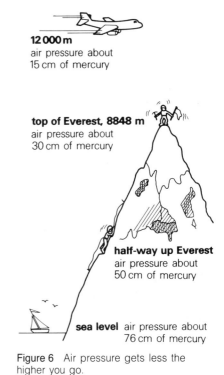

top of Everest, 8848 m
air pressure about 30 cm of mercury

half-way up Everest
air pressure about 50 cm of mercury

sea level air pressure about 76 cm of mercury

Figure 6   Air pressure gets less the higher you go.

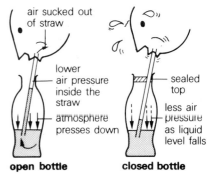

air sucked out of straw

lower air pressure inside the straw

atmosphere presses down

open bottle

sealed top

less air pressure as liquid level falls

closed bottle

Figure 7   Air pressure forces the liquid up the straw.

air pressure

air pressure

milk flows

Figure 8   Two holes are needed before the milk will flow.

Suckers Rubber suckers and suction pads use atmospheric pressure. When you push a rubber sucker onto a flat surface you squeeze most of the air out from under it, as Figure 9 shows. The pressure of the air outside it holds the sucker firmly against the surface.

Pumps Atmospheric pressure is used in pumps of all kinds: bicycle pumps, pumps for pouring beer and raising water, and syringes for giving injections.

Gas pressure

As you know, every gas contains tiny particles called molecules which move around quickly in all directions. As these molecules move around they bump into the walls of their container and push against it. This is what causes *gas pressure*. The moving gas molecules push against anything holding them in. Figure 10 shows a U-tube full of water connected to a gas tap. When the tap is turned on the gas molecules push out in every direction. Some gas molecules push against the water in the U-tube, and force it around.

Manometers The higher the gas pressure the further the water goes around the tube. U-tubes with liquid in, like these, are used to measure gas pressure. They are called *manometers*. Figure 11 shows a large manometer being used to measure the pressure of someone's lungs. As you blow on one side the atmosphere pushes down on the other. The difference in the water levels tells you how much stronger your lung pressure is than atmospheric pressure. All manometers *compare* gas pressure with atmospheric pressure.

With really high gas pressures mercury is used instead of water. It is much more dense so it is not 'pushed around' as much as water.

Bourdon gauge The toy shown in Figure 12 is often used at Christmas parties. The harder you blow the more it uncurls. One instrument for measuring pressure, such as the pressure inside a car tyre, works in the same way. It is called a *Bourdon gauge*. Like the Christmas toy this gauge has a tube that moves when gas is blown into it. The higher the pressure the more the tube starts to straighten out. As the tube uncurls it moves a pointer across a scale. This scale tells you the pressure, usually in N/m².

Large and small pressures

The table below shows you some pressures of different sizes:

	pressure in N/m² (roughly)	pressure in bar (roughly)
the 'normal' atmosphere	100 000	1 bar
inside a car tyre	200 000	2 bar
from an elephant's foot	200 000	2 bar
diver under 30 m of water	300 000	3 bar
about 900 m up Everest	40 000	$\frac{2}{5}$ bar (0.4 bar)
inside an aneroid barometer	20 000	$\frac{1}{5}$ bar (0.2 bar)

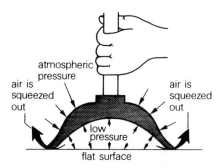

Figure 9 Air pressure holds the sucker firmly on the surface.

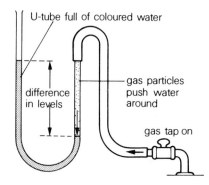

Figure 10 Gas pressure pushes the water around the tube.

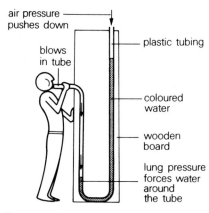

Figure 11 A manometer can compare lung pressure with air pressure.

Figure 12 The harder you blow, the more the toy uncurls.

The atmosphere pushes on everything on Earth with a force of about 100 000 N on every square metre. Figure 13 shows a table top with an area of 1 m². The total force pressing down on it must be about 100 000 newtons, which is roughly the weight of ten average elephants! Why doesn't the table collapse? Luckily there is a good reason.

The Particle Theory explains that air pressure is a result of molecules moving around in all directions, bumping into and pushing against things. 1 cm³ of air contains 10 000 000 000 000 000 000 molecules. So millions of air molecules are bumping into the table top every second. But just as many molecules are striking the table underneath. The air pressures above and below the table are the same.

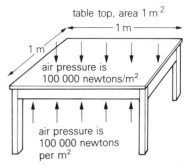

table top, area 1 m²

air pressure is 100 000 newtons/m²

air pressure is 100 000 newtons per m²

Figure 13 The table doesn't collapse because the air pressures above and below it are the same.

Summary

1 You live at the bottom of a 'sea' of air. The pressure of this air is called atmospheric pressure.

2 You cannot feel this pressure – but you can see some of the things it does.

3 Atmospheric pressure is measured by barometers. Some contain a liquid, either mercury or water. Others have no liquid and are called aneroid barometers.

4 You use atmospheric pressure whenever you drink through a straw or pour a drink from a can.

5 All gases have a pressure because their molecules move around quickly in every direction. Gas pressures can be measured with a manometer or a Bourdon gauge.

6 The atmosphere's pressure is about 100 000 N/m² on Earth. The pressure inside a car tyre may be twice as large as this.

This oil pressure gauge is a Bourdon gauge. It is used for measuring oil pressure in cars.

Exercises

1 Where does atmospheric pressure come from? Why does it get less and less as you rise higher and higher? How does it change if you go down a deep coal mine?

2 Describe two ways of showing that the air around you presses on everything.

3 How can atmospheric pressure be measured? Draw simple diagrams of the meters needed.

4 What is 'normal atmospheric pressure'? What units can it be measured in?

5 Explain why:
a) a person can drink through a straw
b) a tin of cream pours more easily with two holes in it instead of one

c) a rubber sucker sticks to a sheet of glass
d) the boiling point of water changes as you climb a high mountain.

6 Describe, with a simple diagram, how a large manometer can be used to measure your lung pressure.

7 Write down the sizes, in N/m², of some large and small pressures.

8 Explain, in your own words, how a Bourdon gauge works. What is it used for?

9 Why don't tables collapse from the huge pressure of the atmosphere above them?

3.8 Floating and sinking

Why does a piece of wood float in water while a piece of steel sinks? Why does a steel ship float in the sea? How does a hot air balloon rise in the sky? The answers depend upon two important ideas in physics: density and upthrust.

Density

Volume 1 kg of polystyrene takes up more space than 1 kg of steel. They both have the same mass, but the polystyrene has a larger volume. The one kilogram of steel is packed into a much smaller space – the steel is more *dense*.

Measuring density The best way to measure the density of a material is to find the mass of 1 m³ or 1 cm³ of it (Figure 1). But materials, especially gases and liquids, are never found in nice, convenient cubes with a volume of exactly 1 m³ or 1 cm³. To find their density you need to measure their *mass* and their *volume*. Suppose a piece of wood has a volume of 2 m³ and a mass of 1600 kg. Then 1 m³ of wood will have a mass of 800 kg. You divide the mass of wood by its volume to find the density:

density	=	mass	÷	volume
(in kg/m³)		(in kg)		(in m³)

Density is measured in kilograms per cubic metre (kg/m³ for short) or sometimes grams per cubic centimetre (g/cm³).

You can find the mass of something by placing it on a lever-arm balance, shown in Figure 2. This balance gives the same reading on the moon, the Earth, or any planet. It tells you the mass of an object in grams (g) or kilograms (kg).

Unusual shapes How can you find the volume of something? One easy way is to lower it into a measuring cylinder, as Figure 3 shows. As the stone goes into the water, the level goes up. The stone pushes water aside – it *displaces* water. If the level goes up by 10 cm³ this must be the space taken by the stone. The stone's volume is 10 cm³. If the stone's mass is 30 g,

its density = $\dfrac{\text{mass}}{\text{volume}}$ = $\dfrac{30\ g}{10\ cm^3}$ = 3 g/cm³.

The density of the stone is 3 grams per cubic centimetre.

Density and floating

It so happens that cold water has a density of one gram per cm³. Figure 4 shows some blocks or cubes made of different materials in a tank full of water. All the materials with a density of more than one gram per cm³ will *sink*. The materials with a density less than 1 g/cm³ all *float*. By putting different materials in water you can

material	mass of 1 m³
air	1.5 kg
feathers	45 kg
wood	750 kg
water	1000 kg
lead	11 300 kg
gold	19 000 kg

Figure 1 The density of a material is the same as the mass of 1 m³ of it.

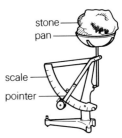

Figure 2 A lever-arm balance shows the mass of an object.

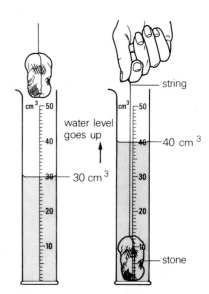

Figure 3 The volume of the stone is the same as the increase in the water level.

76

compare their density with the density of water. Steel has a density of about 8 g/cm³. It is 8 times more dense than water.

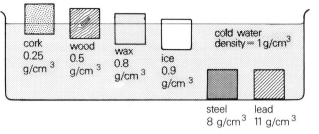

Figure 4 The cubes of steel and lead sink because their density is greater than the density of water.

Figure 5 shows the density of some different 'floaters' and 'sinkers'.

All the floaters have a density less than one g/cm³. So why does a steel ship float?

Upthrust

Whether an object floats or not depends upon *two* things:
- its *density*
- its *shape*.

A steel ship is shaped so that it 'pushes aside' a lot of water. The water pushed aside presses *upwards* on the bottom of the ship (Figure 6). This upward push is called an **upthrust**. The more water the ship pushes aside, the larger the upthrust.

Archimedes' law. Over 2000 years ago a Greek called Archimedes discovered an important law about upthrust:

> the upthrust = the weight of liquid or gas pushed aside

The law applies to all objects in a liquid or a gas, whether they float or not. The stone in Figure 7 'feels' an upthrust when it is lowered into water. It pushes aside 1 N of water, so the upthrust on it is 1 N. But this is not strong enough to hold the stone up. The stone sinks because its weight (2 N) is stronger than the upthrust on it (1 N).

Floating Things float when the upthrust on them is strong enough to balance their weight. As the ship in Figure 6 is lowered into water it pushes more and more water aside, and the upthrust from the water gets stronger and stronger. Eventually, the weight of the ship = the upthrust on it. The ship floats. If the ship weighs 50 million newtons the weight of water it pushes aside must be 50 million newtons too.

Floating in different liquids

Loading a ship A ship floats when the upthrust, pushing it upwards, is strong enough to balance its weight, pulling it downwards. If a ship is loaded with cargo it floats lower down in the water. As it gets lower it pushes more water aside making extra upthrust. This extra upthrust balances the extra weight of the cargo. In the 19th century many greedy shipowners overloaded their ships

floaters	density in g/cm³	sinkers	density in g/cm³
ice	0.9	gold	19
oil	0.8	mercury	14
wax	0.8	lead	11
wood	0.5	brass	8.5
cork	0.25	steel	8
polystyrene	0.01		

Figure 5 The density of some 'floaters' and 'sinkers'

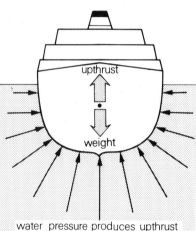

Figure 6 A ship floats when its weight is balanced by the upthrust from the water pushed aside.

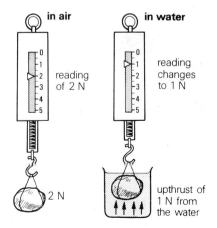

Figure 7 The stone will sink because its weight is stronger than the upthrust on it.

until they floated so low in the water that they sank in a storm. After 1876, Plimsoll lines were marked on the side of every ship. It is against the law to load a ship so that water comes above this line.

Different Plimsoll lines   A ship will float at different levels in different kinds of water. Figure 8 shows some different Plimsoll lines on the side of a ship. 'F' stands for fresh water. The ship floats lower in fresh water than it does in sea-water because sea-water is more dense. As water gets warmer it expands and takes up more space, so warm water is less dense than cold water. 'W' stands for winter sea-water, 'S' stands for summer sea-water. The ship floats lower in summer than it does in winter. It floats lowest of all in 'TF', tropical fresh water, which is quite warm. 'T' stands for tropical and 'WNA' stands for winter North Atlantic.

Hydrometers   1 kg of warm water takes up more space than 1 kg of cold water – it is less dense. Liquids like meths and paraffin are not as dense as water. Figure 9 shows how you can compare liquids with different densities. An ordinary test tube is loaded with lead shot until it floats upright, like a fishing float. The test tube floats at different levels in different liquids.

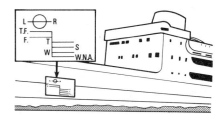

Figure 8   A ship is loaded to float at different levels in different kinds of water.

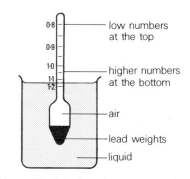

Figure 10   A hydrometer measures the density of a liquid.

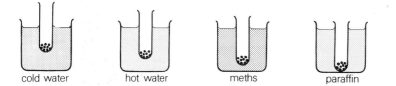

Figure 9   The test tube floats at different levels in different liquids.

Floats like these for comparing the density of liquids are called *hydrometers*. The *lower* they float in a liquid the *less* dense it is. Figure 10 shows a specially made hydrometer with a scale on it. The higher densities are at the bottom of the scale, the lower ones at the top. Hydrometers are used for measuring the density of beer, milk, acid from car batteries, and many other liquids.

Using upthrust

Balloons and airships   When you go swimming or lie in the bath you can feel the upthrust from the water. But even when you are standing up your body is pushing *air* aside – there is a small upthrust on you, but only about $\frac{1}{2}$ N so you do not notice it. Balloons and airships use this upthrust from the air to make them float. They are filled with a gas, like hydrogen or helium, which is about 14 times less dense than air. The upthrust on the balloon is strong enough to make it rise into the air. The hot gas inside the balloon is less dense than the colder air outside. The balloon floats in air, just as wood floats in water.

Submarines   A submarine has special tanks which can be filled with water to make it sink, as shown in Figure 11. To rise to the surface again, the submarine uses upthrust. Water is forced out of the tanks and air is forced in. The upthrust on the submarine is now stronger than its weight and it rises to the surface.

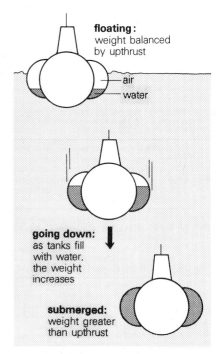

Figure 11   The submarine sinks when its weight is greater than the upthrust.

Weights in water You can use upthrust to lift a heavy rock in water. The rock seems to weigh much less in the water than it does out of it. Fishermen use landing nets to lift a big fish out of the water. Their fishing line might hold the fish inside the water with the help of upthrust but it could break when the fish is in the air.

Floating in water Even a heavy man can use upthrust to help him float in water. If most of his body is in the water a lot of water is pushed aside. The upthrust is then strong enough to keep him afloat. The man in distress in Figure 12 has much less of his body under water – his arms are in the air. There is not enough upthrust to keep him afloat.

One inland sea in Israel has so much salt in it that people can stay afloat reading a book. The water in the Dead Sea is much more dense than ordinary sea-water, and gives a much stronger upthrust.

Figure 12 In this position there is not enough upthrust on the man to keep him afloat. He should lie with his arms in the water.

A submarine rises to the surface when its weight is less than the upthrust on it.

Summary

1 The density of something can be found by dividing its mass by its volume. Density is measured in kg/m³ or g/cm³. The density of water at 4°C is 1 g/cm³.

2 Materials that are more dense than water will sink in water; less dense materials float.

3 When anything is lowered into water there is an upward force on it called upthrust.
If the upthrust is less than its weight it sinks.
If the upthrust equals its weight it floats.

4 A ship floats at different levels in different kinds of water: fresh water, sea-water, warm water, cold water, etc. The density of different kinds of water and different liquids can be measured with a hydrometer.

5 Upthrust is used by: hot air balloons, airships, submarines, fishermen, and people floating in the Dead Sea.

Exercises

1 Which has the larger mass, 1 kg of polystyrene or 1 kg of steel? Which has the bigger volume? Which is more dense?

2 Write down the mass of:
   a) 1 cm³ of lead   b) 1 cm³ of wood
   c) 1 cm³ of gold   d) 1 cm³ of water.
   Which of these is the most dense? Which will float in water?

3 The diagram shows three blocks, of different materials, in water. What can you say about the density of each material?
   What might A, B, and C be made of?

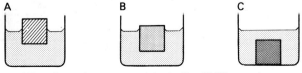

4 What force keeps an object afloat? Where does this force come from?

5 Why do ships have Plimsoll lines marked on the side of them? Explain why more than one line is marked on a ship.

6 Copy out and fill in the blanks:
   The density of a liquid can be measured with a _____ . The lower it floats in the liquid the _____ dense it is. The scale is marked with _____ densities near the top, _____ densities at the bottom. Three liquids with different densities are _____ , _____ , and _____ .

7 Explain why:
   a) a person floats more easily in the Dead Sea than in ordinary sea-water
   b) airships are filled with helium gas
   c) hot air balloons float in the air
   d) a heavy rock seems to weigh less under water than it does in the air
   e) the water level in the bath goes up when you climb in.

Madness or genius?

Many people think of scientists as slightly mad. But is it madness, or genius? Stories are often invented, or the truth is often 'stretched', so that geniuses sometimes seem crazy. Here are three examples, about a Greek, an Italian, and an Englishman. All three spent their lives studying forces and movement.

The world's first streaker

Greek Chronicle 29 February 250 BC

Streaker spotted in city

"A man was seen running naked through the streets of Syracuse yesterday. Shoppers in the pedestrian precinct were shocked by his shouting and arm waving. The man, Mr Archimedes, was promptly arrested and taken to court. When asked the meaning of his strange cry of 'Eureka', the naked man shrieked, 'I've got it, I've got it'.

The judge remarked: 'You may have got it but not everyone wants to see it.' In his defence Mr Archimedes said: 'I was only trying to prove that King Hiero's crown is made of silver and not gold. I found its volume by lowering it under water in the bath. I knew its mass. So I calculated its density – it cannot be gold. The Court goldsmith is the guilty one, not me.'

The judge, in a witty mood, replied: 'This scientific nonsense is all Greek to me. You are fined 50 drachmas and forbidden to experiment for one month.'

King Hiero was incensed, and angry too, when he heard the news: 'Mr Archimedes is one of the top scientists in Greece. He invented the screw, built our latest war engine, and even found π (whatever that is).' The King immediately displaced the Court goldsmith, put him in prison, and ordered him to recite Archimedes' latest law twice daily: 'When an object is immersed in a fluid it experiences an upthrust equal to the weight of fluid displaced.'"

The Italian who dropped cannonballs

Galileo Galilei was born in Pisa, Italy in the same year that William Shakespeare was born in England: 1564. Shakespeare's plays are still read and performed over 400 years later. Galileo's discoveries and experiments are still talked about today.

Galileo is often called the first real scientist. He did experiments. He tested theories. When he

The cathedral and Leaning Tower of Pisa

was a boy people believed heavy objects fell more quickly than light ones. Galileo tested that belief with a simple experiment. He climbed to the top of the Leaning Tower of Pisa. He took two iron cannonballs with him. One had a mass of 5 kg, the other only ½ kg. Next, Galileo made sure that nobody was underneath. Then he dropped the cannonballs.

They both reached the ground at the same time. Galileo had proved that the old theory was wrong. Luckily nobody was watching him – they might have thought he was mad.

Galileo made another of his discoveries while he was sitting in the Cathedral, next door to the Leaning Tower. A lamp, hanging from the Cathedral roof, was swinging to and fro. Instead of listening to the priest's sermon Galileo watched the lamp. He saw that the time for each swing was always the same. Years later, as an old man, he designed a pendulum clock.

But unfortunately nobody believed Galileo's ideas and experiments. He even had the cheek to suggest that the Earth went round the Sun . . .

Seeing stars Galileo was the first man, in 1609, to make discoveries with a new-fangled Dutch invention: the *telescope*. He was the first to spot the mountains and craters on the moon. He saw many stars never seen before, and the galaxy we now call the Milky Way. He even discovered four of Jupiter's moons.

This made Galileo famous all over Europe. He soon became a Professor at Pisa. But it also got him into big trouble with THE CHURCH. Galileo, now famous, was proclaiming: 'The Earth is *not* the centre of the Universe – it moves.' The

Church offered Galileo a choice: apologise or have your head cut off. Galileo was a clever man. He apologised. But as he was dragged away he mumbled: 'The Earth does move.'

Just to be safe the Church kept him in his own home for the rest of his life.

Another new idea   The ancient Greeks believed that whenever something moves a force must be pushing it: when an arrow flies through the air, air from in front must be rushing around and pushing the arrow from behind. Galileo helped to disprove this theory and propose a new one: 'Once something is moving it keeps moving, until a force stops it.' Nobody really believed him. How could something keep moving if it wasn't pushed? How did the planets keep going?

Galileo's ideas were only accepted after he died. He was finally forgiven by the Church of Rome in October 1980.

Falling apples in autumn

In the same year that Galileo died the greatest physicist of all time was born: Isaac Newton arrived on Christmas Day in 1642, on a farm in Lincolnshire.

His father died before he was born. When he was three his mother married a vicar and left young Isaac with his grandmother. Isaac's upset early childhood probably affected him for the rest of his life. He was always known as an 'absent-minded Professor'.

Newton was not much good as a farmer. Most of the farm workers were pleased when he was sent off to Cambridge University. He didn't do very well at Cambridge either. Then, in 1665, the Plague hit England. Isaac Newton went back to Lincolnshire to escape from it. In the next two years Newton made nearly all the discoveries that have made him famous.

Once upon a time, on a warm autumn afternoon, Newton was sitting in his orchard. Suddenly, an apple fell on his head. 'Ouch', said Isaac, cursing to himself. But the blow set him thinking about *gravity*. Why did the apple fall? Newton reasoned that *every* object attracts *every other* object towards it with the pull of gravity. The force is *universal*; gravity is everywhere! The Earth's gravity extends far out into space, far above the highest apple tree. The Sun's gravity extends through the solar system. It is the force of the Sun's gravity which keeps the planets in their orbits. This was Newton's greatest 'discovery'.

Still avoiding the Plague, Newton made the first *spectrum* of colours. He shone sunlight through a specially shaped piece of glass called a prism. Newton realised that white light is no more than the seven colours of the spectrum put together.

When it was too wet to go in the orchard, and not sunny enough to use a prism, Newton did mathematics. He devised 'the Calculus'. (It still drives sixth-formers mad 300 years later.)

After these discoveries Newton became the new Professor at Cambridge. At the age of 30 he made the first *reflecting telescope*. The same kind of telescope is still used by astronomers. But the thing that most people remember about Newton are his three *laws of motion*.

Newton's reflecting telescope

Apart from physics and mathematics Newton had plenty of other interests: alchemy, theology, philosophy, studying ancient civilisations, coinage. In 1700 he even became Master of the Royal Mint. He never got married, probably because he didn't have time. After he died, in 1726, he was given a national funeral at Westminster Abbey.

It makes you wonder what science laboratories are for: Archimedes made his greatest discovery lying in the bath. Galileo made his famous discoveries in the Leaning Tower of Pisa and the Cathedral next door. Newton had his greatest thoughts, according to a well-known joke, after an apple fell on his head. It just goes to show that scientific discoveries can happen anywhere, if there are geniuses about.

Topic 3 Exercises

More questions on forces

1 This aircraft is flying *straight* and *level* at a steady
   speed. Four forces are shown acting on it:
   a) What force is pushing it forwards?
   b) What force is acting in the opposite direction?
   c) What can you say about these two forces?
   d) What can you say about the forces acting
      upwards and downwards?

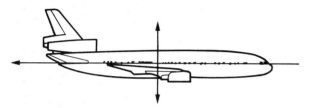

2 Here are six diagrams of bricks resting on a
   bench. Some are resting on a nail, or two nails.
   List these in order with the greatest pressure on
   the bench first, lowest pressure last.

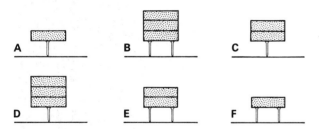

3 The left diagram below is of a mercury barometer.
   a) What is there at **A**?
   b) What is the height of the column on a 'normal'
      day?
   c) Explain how the barometer works.
   d) What is the other type of barometer called?

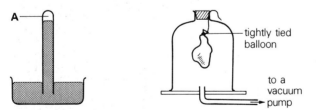

4 The diagram on the right above shows a partly
   blown-up balloon inside a large glass jar (called a
   bell jar). What happens to the balloon as the air is
   sucked out of the jar?

5 What would happen to the volume of a balloon if it
   was inflated on the surface of the sea and then
   taken down to about 10 metres below the surface?
   and vice versa?

Things to do

1 Rolling uphill?

   You need: thick paper, scissors, sellotape, two
   straight sticks, a large and a small book.
   Make two cones from the paper then join them
   together. Rest the two sticks across the books.
   Then place the cone on the two sticks, next to the
   small book. It appears to roll uphill.

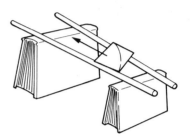

   In fact the cone does not defy gravity. If you look
   closely the *points* of the cones really go
   downwards.

2 Balancing a potato

   You need: a potato, two forks, an egg cup, and a
   2p coin. Stick the forks and the coin in the potato
   as the diagram shows. Hold the egg cup. The
   potato will now rest on the coin on the very edge
   of the egg cup.

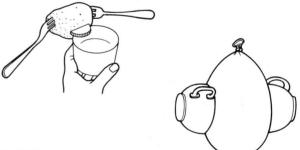

3 Picking up tea-cups

   You need: a balloon, two *old* tea-cups. Blow up the
   balloon slightly and press the tea-cups against it.
   Then blow the balloon up completely. You can now
   lift the balloon up, and the cups with it . . .

4 Air supporting water

   Completely fill a jam jar with water. Cover it with a
   thin card. Hold the card tightly on the jar, and then
   turn the jar upside down.

   Take your hand away. The water does not fall out.
   The air pressure below the card is easily strong
   enough to support the weight of the water above.

Waves

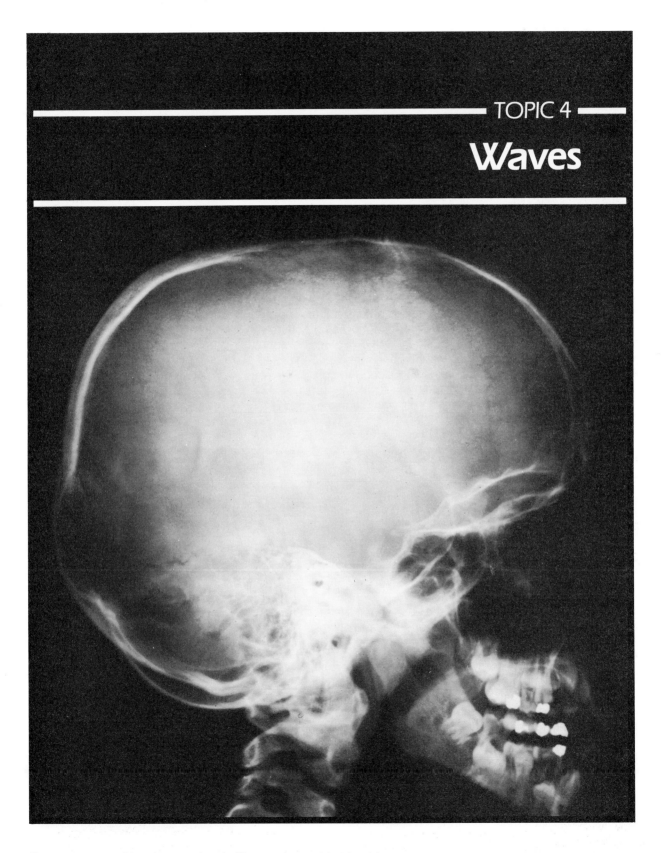

X-rays are waves with a short wavelength. They can be used to take pictures inside a human body.

4.1 What is a wave?

What do lasers and ripples on a pond have in common? How do heat and light reach us from the Sun? The answers depend upon 'energy carriers' called waves.

Different waves

Making waves If a stone is dropped into the middle of a very still pond it makes ripples. These ripples spread out from the stone until they reach the edge of the pond, as Figure 1 shows. A ball floating at the edge of the pond moves up and down as the ripples move past it. Some of the energy of the stone is given to the ball. Waves on the sea are like ripples on a pond – as they move *across* the sea a boat floating on it moves *up and down* (Figure 2). You can make a wave by tying one end of a rope to a post and moving the other end up and down. Each part of the rope moves up and down as the wave moves along the rope. Figure 3 shows a special spring called a 'slinky spring'. If you stretch the spring then jerk one hand from side to side this movement is carried along the spring, like a wave. If you use your hand to push the spring in and out, this movement travels from one end to the other, through the spring.

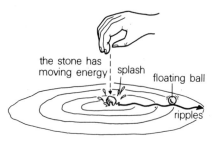

Figure 1 Energy from the stone is carried to the ball by the ripples.

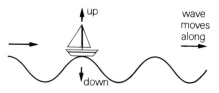

Figure 2 As a wave moves across the water, the boat moves up and down.

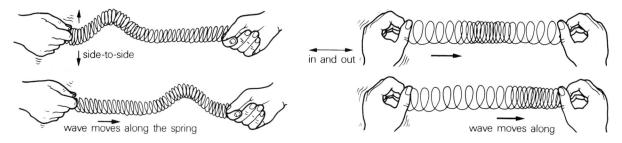

Figure 3 Movement travels along the spring as a wave.

Ripples on a pond, waves on the sea, and waves in ropes and springs all carry moving energy from one place to another. A wave is a way of *carrying energy* from one place to another.

Why study waves? Light rays behave like waves – they carry energy from one place to another. The Sun's rays carry light energy from the Sun to the Earth. Heat rays carry energy too. Heat rays carry heat energy from an electric fire to a cold room. Sound waves carry sound energy from a loudspeaker or a person's voice to your eardrums.

Studying waves helps us to understand more about light, heat, and sound, and how they behave. A special tank for studying waves is called a *ripple tank*. It can be used to show:
- different types of wave
- how waves travel
- how waves can be reflected
- how waves can 'bend'.

A surfboard rider uses the moving energy of a water wave.

Looking at waves

A ripple tank is just a see-through dish with water in it. Figure 4 shows a very simple ripple tank.

Plane and circular waves Look at Figure 5. If you move a wooden bar up and down in the tank it makes straight waves, called *plane* waves. But if you dip your finger into the water, ripples spread out from it in circles. These are called *circular* waves. Circular waves are like ripples on a pond.

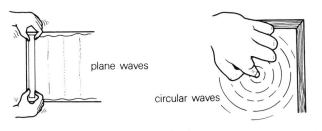

Figure 5 Waves can be plane waves or circular waves.

Wavelengths Plane and circular waves both have high parts, called *crests*, and low parts, called *troughs*. You can see these crests and troughs if you look at the wave from the side, as Figure 6 shows. If the wooden bar moves up and down quickly it makes more waves every second – the distance between crests gets smaller and smaller. This distance is called the *wavelength*. When the bar moves up and down slowly the wavelength is long – when it moves quickly, the wavelength is short, as Figure 7 shows.

Some waves are plane waves, some are circular waves. All waves have different wavelengths. The ripple tank is used to study these different types of waves.

How do waves behave?

Reflecting waves When waves in a ripple tank meet a straight barrier they bounce off it, as Figure 8 shows. The ripples are *reflected*. You can see that the reflected waves leave the barrier at the same angle as the waves that hit it. Figure 9 shows what happens when plane waves meet a curved barrier – the waves bounce off the barrier and travel back as circular waves towards one point. This point is called the *focus*. The curved barrier can also change circular waves to plane waves. If you dip your finger in the water at the focus it sends out circular waves. The curved barrier changes them into plane waves, as Figure 10 shows.

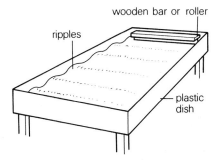

Figure 4 A simple ripple tank

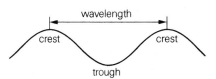

Figure 6 The side view of a wave shows the crests and troughs. The distance between the crests is called the wavelength.

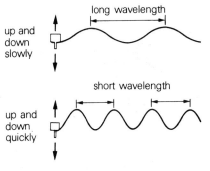

Figure 7 Shorter waves are made by moving the bar more quickly.

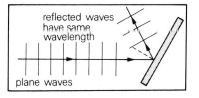

Figure 8 Plane waves are reflected from a straight barrier as plane waves.

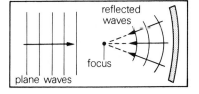

Figure 9 Plane waves are reflected from a curved barrier as circular waves.

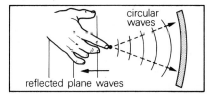

Figure 10 Circular waves are reflected from a curved barrier as plane waves.

If the curved barrier is turned the other way round it makes a plane wave 'spread out', as Figure 11 shows.

Straight and curved barriers are used in a ripple tank to show how waves are reflected.

Bending waves   You can also use a ripple tank to show how waves can be bent or *refracted*. Figure 12 shows a specially shaped piece of plastic in the ripple tank. The water above this plastic is not as deep as the water in the rest of the tank. When a plane wave travels from the deep water to the shallow water it starts to bend. The plane wave is refracted as it meets the edge of the plastic.

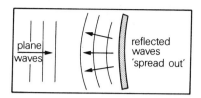

Figure 11   If plane waves are reflected from the inside of a curved barrier, they travel towards one point. But if they are reflected from the outside of a curved barrier, they spread out.

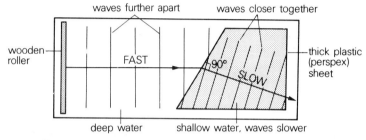

Figure 12   The bending of a wave is called refraction.

All waves can be reflected and refracted. These two important properties of waves will be studied closely in the rest of this topic.

Different types of wave

Fast and slow   Some waves travel very quickly, others travel quite slowly. In a thunderstorm you see the lightning flash before you hear the thunder. Light travels much more quickly than sound.

Something to travel in   Some waves can travel through empty space. Light and heat rays from the Sun travel about 150 million kilometres before they reach the Earth. Most of their journey is through empty space – a vacuum. But sound waves need something to travel in. Figure 13 shows a bell ringing inside a large glass jar. When the air is removed from the jar the bell goes quiet. You can see the bell ringing but you cannot hear it.

Sound can travel in water, air, metal, brick, or wood, but never in a vacuum.

Long and short wavelengths   Sound waves can have a wavelength as small as one-hundredth of a metre, or as large as two metres.

Some waves have very long wavelengths. Radio waves can have wavelengths of 275 m, 285 m, or even 1500 m. Other wavelengths are very short. Light wavelengths are only about one-millionth of a metre.

A special family of waves

Some waves are slow, some fast; some can travel in a vacuum, some can't; some have short wavelengths, others long. This could get very confusing. Luckily there is a special family of waves in physics which have certain things in common.

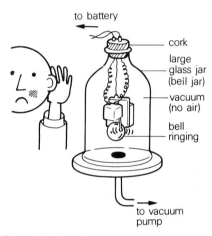

Figure 13   Sound waves cannot travel through a vacuum.

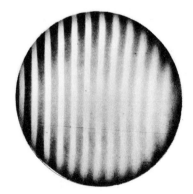

This pattern in a ripple tank is made by plane waves.

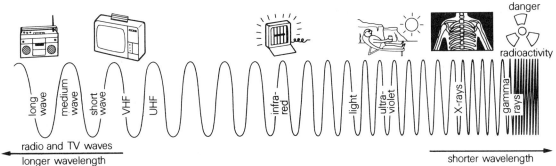

radio and TV waves
longer wavelength

shorter wavelength

Figure 14 The electromagnetic spectrum is a family of waves. They all have different wavelengths, but they all travel at the same speed as light, and they can all travel through a vacuum.

This family or group contains all the waves which:
■ travel at the same speed as light – a speed of about 300 000 000 metres per second
■ can travel through empty space (a vacuum).

This family of waves is called the *electromagnetic spectrum*. Look at Figure 14. All the waves have different wavelengths. Radio waves have the longest wavelengths. X-rays and gamma rays have very, very short wavelengths – about one million-millionth of a metre! Inbetween are infra-red rays, light rays, and ultra-violet rays. Ultra-violet rays are the rays from the Sun that give you a sun tan.

Most of this topic tells you how light waves can be reflected, refracted, and used to help you see small things and things a long way off. One important wave does not belong to 'the family' – the sound wave. Sound waves will be studied at the end of the topic.

This 'sun-lamp' gives out both ultra-violet and infra-red rays. The ultra-violet rays can be used to get a suntan. The infra-red rays are used to relieve muscular pain.

Summary

1 A wave is one way of carrying energy from one place to another.

2 Waves can be studied in a ripple tank.

3 All waves can be reflected and refracted.

4 Many different waves are studied in physics: some are slow, some fast; some need a material to travel in, some can travel in empty space; some have long wavelengths, others short.

5 Many waves belong to a family called the electromagnetic spectrum: they can all travel through empty space (a vacuum) at the same speed (300 million metres per second).

Exercises

1 Write down three different examples of waves.

2 What is a ripple tank used for? What kinds of wave can be made in a ripple tank?

3 Describe carefully how waves in a ripple tank can be: a) reflected   b) refracted.

4 Write down three materials that sound waves can travel in.

5 Which waves can travel through empty space?

6 Give two examples of waves with:
a) very short wavelengths   b) long wavelengths.

7 What is special about waves in the family called the electromagnetic spectrum?

8 Draw this table and complete it to show 6 different waves in the electromagnetic spectrum by labelling A, B, C, D, E, and F:

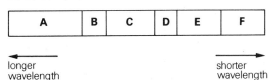

A	B	C	D	E	F

← longer wavelength

shorter wavelength →

4.2 How do light waves travel?

How can you see the moon even though it does not shine? How are shadows made? What causes eclipses? The answers to these questions all depend upon the way that light waves travel.

Where do waves come from?

Waves are often called rays. Whenever something gets hot it gives off heat rays. If it becomes red-hot and starts to glow it gives off light rays. Anything which gives off rays or waves is called a *source*.

Common sources   The Sun, an electric fire, or a red-hot nail are sources of heat rays. A bell, a loudspeaker, or a person's voice are sources of sound waves. Radio transmitters are sources of radio waves. You will see in Topic 5 that 'excited atoms' are sources of X-rays and gamma rays. All waves come from a source.

Luminous sources   Anything which makes light waves is called a *luminous* source. The Sun, other stars, light bulbs, television screens all give off their own light – they are all luminous.

Whenever you look at a luminous object, light rays from it travel *directly* into your eyes. This is how you see them. But most objects do not give off their own light. This page, other people, the moon, cat's eyes in the road, most everyday objects are non-luminous. They do not make light themselves. You see them when they **reflect** light from a luminous source into your eyes, as Figure 1 shows.

Rays and beams

Most luminous sources give off light in every direction. The light and heat rays from a light bulb spread out in all directions as Figure 2 shows, just like ripples on a pond.

Whenever you see light rays you see them collected together as a *beam*. Figure 3 shows three different types of beam: *diverging, parallel*, and *converging*. The light from most luminous objects is

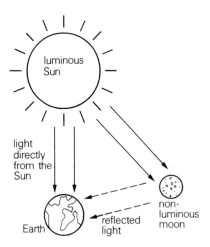

Figure 1   The Sun gives off its own light. The moon does not give off its own light. You can see the moon because it reflects light from the Sun.

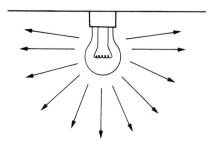

Figure 2   Light and heat rays from a bulb spread out in all directions.

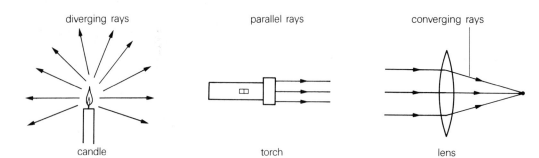

Figure 3   A beam of light rays can be diverging, parallel, or converging.

diverging. Mirrors can be used to make this light into a parallel beam. Torches, searchlights, and car headlights have a special mirror behind the bulb to make a parallel beam. The shiny reflector behind an electric fire makes an almost parallel beam of heat rays. Lenses are used to make a converging beam of light.

Making shadows

Shadows are made because:
- light rays travel in straight lines, and they cannot bend around corners
- some things will not let light go through them.

You can use your hands to stop light rays, and make shadows of different shapes. Look at Figure 4. The shape of the shadow is the same as the outline of the hand. This shows that the light rays are travelling in straight lines.

Types of shadow   There are two types of shadow called the *umbra* and the *penumbra*. The umbra shadow is completely dark. If you hold a football in front of a small light bulb, you can see the umbra type of shadow on a screen or a white wall. As Figure 5 shows, *none* of the light from the bulb has reached the umbra.

Figure 4   Light travels in straight lines. This means the shape of a shadow is the same as the shape of its object.

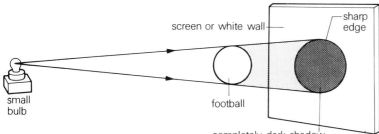
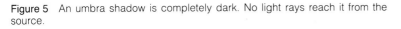

Figure 5   An umbra shadow is completely dark. No light rays reach it from the source.

With a large bulb the football makes two kinds of shadow: very black at the centre, the umbra, and grey around the edges, the penumbra. *Some* of the light from the bulb is reaching this grey area of shadow, as Figure 6 shows.

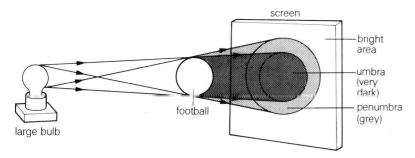
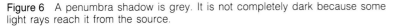

Figure 6   A penumbra shadow is grey. It is not completely dark because some light rays reach it from the source.

Light travels in straight lines, so the shape of the shadow is the same as the shape of the object.

Eclipses

In August 1999 you may be lucky enough to see an eclipse of the Sun in Britain. Eclipses happen when the moon comes between the Sun and the Earth and throws a huge shadow across the Sun. The moon makes the two types of shadow, penumbra and umbra, as Figure 7 shows.

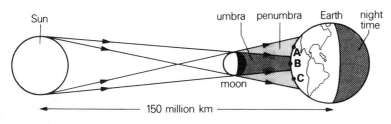

Figure 7 During an eclipse, the moon makes both an umbra and a penumbra shadow.

Total and partial eclipses Anyone standing in the umbra part of the moon's shadow witnesses a *total eclipse*. All the sun's rays are stopped as the umbra part of the moon's shadow, only a few kilometres wide, travels over the Earth. Only the flames on the outer edge of the Sun can be seen. A girl in the penumbra region of the moon's shadow can see part of the Sun, with a shadow across the rest of it. She sees a *partial eclipse*. Figure 8 shows how the Sun looks from different parts of the Earth during an eclipse.

What do eclipses tell you? During a total eclipse the Earth suddenly becomes cold as well as dark. This tells you two things:
- the Sun's rays travel in straight lines – this is why a total eclipse begins so suddenly. The umbra has a sharp, definite edge.
- the heat and light from the Sun travel together at the same speed: 300 million metres per second. Heat and light rays both take about 8 minutes to travel the 150 million kilometres from Sun to Earth. The Sun could have gone out seven minutes ago!

Pinhole cameras

A simple type of camera uses the fact that light travels in straight lines. It can be made from a closed, light-proof box with a tiny pinhole at the front, as Figure 9 shows. Light rays from the object

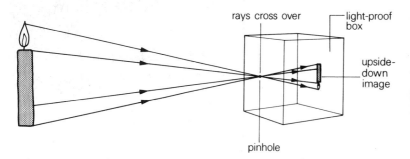

Figure 9 In a pinhole camera, light rays from an object cross over at the pinhole and make an upside-down image at the back of the camera. A large walk-in version of the pinhole camera is called a *camera obscura*.

view of the Sun from **A** in Figure 7

partial eclipse

view of the Sun from **B** in Figure 7

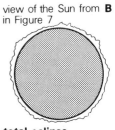

total eclipse

view of the Sun from **C** in Figure 7

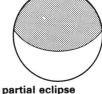

partial eclipse

Figure 8 People at **A**, **B**, and **C** in Figure 7 have different views of the Sun during an eclipse.

In this total eclipse of the Sun, only the flames on the outer edge of the Sun can be seen.

being photographed cross over at the pinhole and make an upside-down *image* at the back of the camera. This image is very faint because the pinhole hardly lets any light in. Modern cameras use a much bigger hole with a special lens to make an image at the back of the camera.

Your eyes have a tiny 'hole' at the front, just like the pinhole camera. Whenever you see something light rays enter your eye through this hole, called the *pupil*.

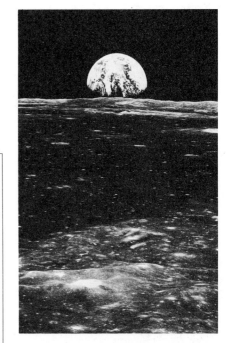

This photo was taken on the moon. The Earth looks bright because light rays from the Sun are reflected from it.

Summary

1   All waves travel outwards from a source. Luminous sources give off their own light waves.

2   You see something when light rays from it enter your eyes.

3   Light travels in rays and beams. There are three types of beam: diverging, parallel, and converging.

4   Shadows are made because light travels in straight lines. There are two kinds of shadow: umbra (completely dark) and penumbra (partly dark).

5   You can see these two types of shadow during an eclipse. The eclipse is total in the umbra region of the moon's shadow, partial in the penumbra.

6   Pinhole cameras use the fact that light travels in straight lines.

Exercises

1   Copy out and fill in the missing words:
Anything which gives off waves is called a _____
A bell, a voice, and a loudspeaker are _____ of waves. Something which gives off light waves is called a _____ _____ . But most objects you see every day are non-_____ .

2   Write down three sources of heat rays, and three sources of light rays.

3   Copy out and fill in the missing words:
The Sun's rays travel in _____ lines, at a speed of _____ metres per second. They take _____ minutes to reach the Earth from the Sun, a distance of _____ kilometres.

4   Use a diagram to explain how eclipses are caused. What is the difference between a total and a partial eclipse?

5   Explain how you see:
a) distant stars
b) this page
c) the moon
d) cat's eyes on the road at night

6   What are the three types of beam?

7   a) Why are shadows made? Name the two types of shadow.
   b) What types of shadow can you see in this photograph?
      How do you think these two types of shadow are formed?

8   Draw a simple diagram to show how a pinhole camera works. Explain why the image at the back of the camera is upside down.

4.3 Reflecting waves

When can you see a candle burning underwater? How can you make yourself shrink or grow? How can a man 2 metres tall see over a wall 3 metres high? It's all done by mirrors.

Reflectors

In unit 4.1 you saw that water waves can be reflected in a ripple tank. When they meet a barrier they bounce off it. All rays and waves can be reflected: radio waves, sound waves, infra-red rays, light rays, and so on. There is a layer in the Earth's atmosphere called the ionosphere. Figure 1 shows how it can be used to reflect radio waves from one part of the world to another.

Echoes Whenever sound waves are reflected you hear an echo. Hard, solid surfaces, like a wall or a cliff, are good reflectors of sound waves. If a boy shouts inside a cave he hears more than one echo, as Figure 2 shows.

Heat rays Silvery, shiny surfaces are good reflectors of heat rays, as you saw in Topic 2.

Light rays Every object you see around you is reflecting light rays – if it didn't you would not be able to see it. But the best reflectors of light are smooth, shiny surfaces. These are used to make mirrors, as Figure 3 shows.

Most of this unit tells you about mirrors, how they reflect light, and how they can be useful.

How do mirrors reflect light?

A ray-box You can study how mirrors reflect light by using a *ray-box*, shown in Figure 4. A ray-box uses a thin slit to make a very narrow beam of light. This is about as near as you can get to a single ray. When this narrow beam from a ray-box meets a mirror it bounces off it – the ray is reflected.

Equal angles The ray of light that strikes the mirror is called the *incident* ray. The ray that bounces off the mirror is called the *reflected* ray. When you use a ray-box you find that the incident ray always makes the same angle with the mirror as the reflected ray. This is true whenever a light ray is reflected.

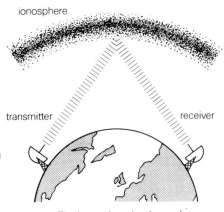

Figure 1 The ionosphere is a layer of charged particles in the Earth's atmosphere which acts as a reflector of radio waves. Satellites are more commonly used today but they are not really reflectors. They absorb radio waves and then re-transmit them.

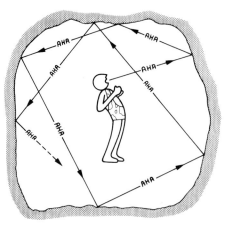

Figure 2 An echo is simply a reflected sound wave.

Figure 3 Smooth shiny surfaces make the best mirrors.

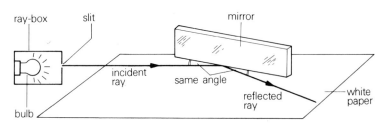

Figure 4 A ray-box is used to study the reflection of light.

You can prove it by shining rays onto a mirror at several different angles, and tracing their paths onto white paper. Then a line at right-angles (90°) to the mirror is drawn in, as Figure 5 shows. This line is called the *normal*. Using a protractor the angle between each ray and the normal can be measured. You find that the angles on both sides of the normal are the same. This is one of the laws of reflection:

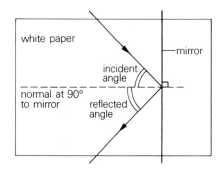

> the incident angle = the reflected angle

The same law applies whenever radio waves, sound waves, heat rays, water waves, or waves of any kind are reflected.

Figure 5 When a ray is reflected, the incident angle is always the same size as the reflected angle.

Your mirror image

Flat, straight mirrors are called *plane* mirrors. Whenever you look into a plane mirror you see a 'picture' of yourself behind the mirror. This is your *mirror image*.

Look at Figure 6. There are four important facts about the picture or image that you see in a plane mirror:

- The image in the mirror is the *same size* as the object in front of it.
- The image is the *right way up*.
- The image is *as far behind* the mirror as the object is in front.
- The image appears *left-to-right* – anything on the left of the object is on the right of the image.

You can test each of these facts about mirror images simply by standing about 2 metres in front of a plane mirror with your right arm raised. Your mirror image appears to have the left arm raised. If you take one large pace forward, about one metre, your mirror image moves the same distance towards you. Step forward again and you and your image will almost be touching!

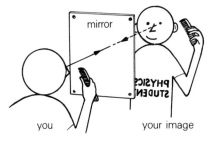

Figure 6 Seeing yourself as you really aren't . . .

Real and virtual images You can see a mirror image. But the image in a plane mirror can never be formed on a screen, like the image of a film can be. The image is called a *virtual* image, meaning that it cannot be focused onto a screen. Images that can be formed on a screen are called *real* images.

Studying the images made by mirrors and lenses is an important part of physics called *optics*. Images can be large or small, the right way up or upside-down, real or virtual. You will meet different types of images all through this topic.

It's all done by mirrors

Plane mirrors can be used to do various 'tricks'. Here are three examples:

Pepper's ghost A sheet of clear glass can act like a mirror, as Figure 7 shows. By putting a beaker full of water exactly where the image of the candle is, it looks as if a candle is burning under water. The image of the burning candle is sometimes called 'Pepper's ghost'.

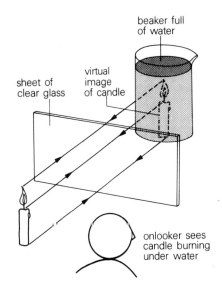

Figure 7 The candle seems to be burning under water.

AMBULANCE

This is often written on the front of ambulances. A driver looking at the word through the rear-view mirror would see: AMBULANCE. Any back-to-front writing appears the right way round in a mirror.

Periscopes   You can use two plane mirrors to see around corners, or over a wall. Look at Figure 8. The top mirror is at an angle of 45° so that it reflects light rays downwards onto the second mirror. This second mirror reflects the rays straight into your eyes. You see a clear image of the object being viewed. This arrangement of two mirrors is called a *periscope*. Long periscopes are used in submarines to see over the surface of the sea.

Curved mirrors

Not all mirrors are straight, plane mirrors. There are two types of curved mirrors, called *convex* and *concave* mirrors.

Convex mirrors   Convex mirrors curve outwards, as Figure 9 shows. A ray-box can be used to shine a parallel beam of light rays onto a convex mirror. When they strike the mirror they are reflected *outwards*. The parallel beam becomes a *diverging* beam that appears to have spread out from a point behind the mirror. The point that the reflected rays seem to have come from is called the *focus*.

Concave mirrors   Concave mirrors curve inwards (remember: 'cave' – goes in). When a beam of parallel light rays hits a concave mirror the rays are reflected *inwards*, as Figure 10 shows. These reflected rays meet, and cross over, at a point called the *focus*. The parallel beam of light is turned into a *converging* beam by this type of mirror.

> A concave mirror is a converging mirror
> A convex mirror is a diverging mirror

Images in curved mirrors   Because of their shapes, curved mirrors make different kinds of images. You can see your image in convex and concave mirrors by looking into both sides of a shiny steel spoon. In convex mirrors the image is always smaller than the object. Sometimes the image in a concave mirror is smaller and upside-down, but when the object is very close to the mirror it makes an image that is bigger than the object itself – this is called a *magnified* image.

Using curved mirrors

Using convex mirrors   Convex mirrors make things look smaller, as Figure 11 shows. This means that you can see much more with a convex mirror than you can see with a plane mirror. A convex mirror gives you a 'wide field of view'. The things you see appear much smaller – but you can see a lot more of them. Convex mirrors are used in:

  car wing mirrors, to see through a wide angle behind the car (look at the photo on page 92),

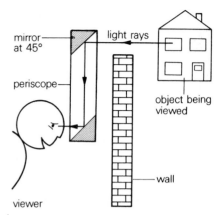

Figure 8   You can see over a wall by using a simple periscope.

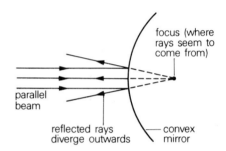

Figure 9   A convex mirror is a diverging mirror.

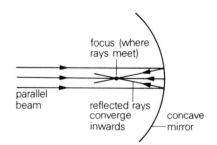

Figure 10   A concave mirror is a converging mirror.

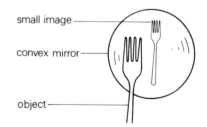

Figure 11   Convex mirrors make things look smaller.

supermarkets and big shops to look out for thieves and shoplifters,

buses, so that the driver can see what the passengers are up to.

Using concave mirrors Concave mirrors can be used to make things look larger, as Figure 12 shows. The image in the mirror is much larger than the object. This makes concave mirrors very useful for shaving or putting on make-up.

Concave mirrors can also be used to reflect heat rays from the Sun so that they are all concentrated at one place. Figure 13 shows a solar power station which uses the Sun's rays to generate electricity.

Figure 14 shows a special type of concave mirror, shaped like the end of a rugby ball. It is called a *parabolic mirror*. It can reflect the rays from a light bulb to make a very straight, parallel beam. Parabolic mirrors are used in car headlights and searchlights, and in electric fires to make a strong, straight beam of heat rays.

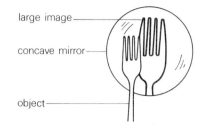
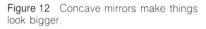

Figure 12 Concave mirrors make things look bigger.

Figure 13 A 'solar power' installation in France.

Summary

1 All rays and waves can be reflected. Hard, shiny surfaces are the best reflectors.

2 Mirrors are used to reflect light rays. One law of reflection says that the incident angle equals the reflected angle.

3 When you see an object in a plane mirror its image is:
a) the same size as the object
b) the wrong way round
c) as far behind the mirror as the object is in front
d) virtual – it cannot be put onto a screen.

4 Plane mirrors can be used in periscopes, and for playing tricks.

5 Curved mirrors are convex or concave.
Convex mirrors are diverging, concave are converging.

6 Convex mirrors are used in supermarkets, buses, and cars.
Concave mirrors are used for shaving and putting on make-up.

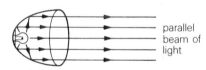

parallel beam of light

parabolic mirror in searchlight

Figure 14 A parabolic mirror makes a parallel beam of light.

Exercises

1 Draw the mirror image of each letter:
F   R   B   C   E   X   Z

2 Describe, with a drawing, how television pictures reach England directly from America.

3 How are echoes made?

4 The drawings show the image of a clock-face in a mirror. What is the right time in each case?

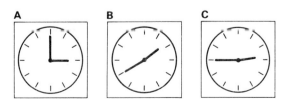

5 Why does a right-handed tennis player look left-handed in a mirror?

6 Draw a rough sketch to show how a submarine captain can see above the surface of the sea, when the submarine is submerged.

7 What is a parabolic mirror used for?

8 Make a table to show the images in plane, convex, and concave mirrors and the uses of each one. Like this:

	plane	concave	convex
image			
uses			

4.4 Bending light waves

Light waves usually travel in straight lines at a steady speed. But they can be bent, and slowed down. This unit tells you how bending light waves can make the bottom of a swimming pool look closer, coins appear and disappear, and small things look bigger.

Refraction

Figure 1 shows a beam of light rays travelling through a glass block. When the rays go *into* the glass they bend one way – as they *leave* the glass block they bend the other way. The light rays bend as they travel from air into glass, and again when they leave the glass and enter the air. But if light meets a surface exactly at right angles it passes straight through, as Figure 2 shows.

The bending of light rays when they travel from one material into another is called *refraction*. People have studied refraction since the days of Ptolemy, about 100 AD. Light rays can be bent, or *refracted*, by water, diamond, glass, plastic, or any material that carries light.

Explaining refraction   Why do light rays bend when they leave one material and enter another? The answer depends upon the *speed* of light. In empty space light travels about 300 million metres every second – its speed is almost as fast in air. But in glass it slows down to about 200 million metres per second. As the light rays slow down they bend.

In unit 4.1 you saw that water waves in a ripple tank are refracted when they travel from deep water to shallow water. The water waves are slowed down, just as light rays are slowed down by glass.

Refraction in water

The speed of light in water is about 225 million metres per second. This means that light rays are slowed down and bent almost as much by water as they are by glass.

The appearing coin   Light rays are bent when they travel in water or when they leave water. You can use this refraction of light to make a coin appear at the bottom of a cup. Stand back until the coin is just out of sight in the empty cup, as Figure 3 shows. Without moving your head, pour water into the cup. As the cup fills with water the coin appears. Light rays from the coin are bent towards your eye as they leave the water.

Bent sticks   For the same reason a straight stick looks bent when it is lowered into some water. The light rays from the bottom of the stick are refracted as they leave the water and enter the air.

'Shallow' pools   Have you ever tried diving for something on the bottom of a swimming pool? It always looks closer than it really is. This is because light rays from the bottom of the pool are refracted as they leave the water. In the same way a glass block can make the letters on a page seem closer than they really are.

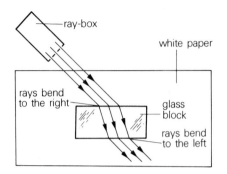

Figure 1   Light rays bend as they travel from air to glass, and from glass to air.

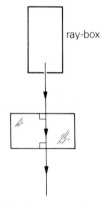

Figure 2   If light meets a surface exactly at right-angles, it is not bent.

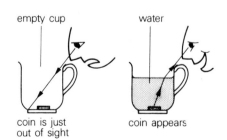

Figure 3   Water refracts the light and the coin becomes visible.

Refraction makes it very difficult to see things under water as they really are. Successful spear-fishers never aim their spear at the place where the fish *appears* to be – the fish is really much deeper (see Figure 4).

Refraction in lenses

Shaping glass Glass can be made into special shapes so that it bends light in certain ways. Figure 5 shows two of these shapes – they are called *lenses*. Lenses that are thicker in the middle than at the edges are called *convex* lenses. Lenses that are thinner in the middle than at the edges are called *concave* lenses.

Not all lenses are made of glass. To read these words you are using a lens inside your eye that is not a glass lens. Any material that carries light can be used to make a lens if it is made into the right shape: diamond, clear plastic, or even water.

Bending light with a lens Figure 6 shows a parallel beam of light rays travelling through a convex lens. The rays are bent or refracted so that they all meet, and cross over, at one special point. This point is called the *principal focus*. On a sunny day you can use a convex lens to focus the Sun's light and heat rays onto a piece of black paper. The point where the rays meet gets burning hot.

A convex lens changes a parallel beam into a converging beam:
A convex lens is a converging lens.
But a concave lens bends a parallel beam of light rays so that they spread out, as Figure 7 shows:
A concave lens is a diverging lens.

Strong and weak lenses Some lenses can be shaped so that they are much stronger than others. Figure 8 shows one weak, and one strong, convex lens. The strong one is much fatter in the middle – it brings light rays to a focus much closer to the lens. Different shaped lenses have different strengths.

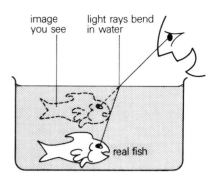

Figure 4 The fish is deeper than it seems.

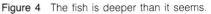

convex lens

concave lens

Figure 5 Side view of two lenses

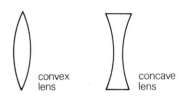

Sun

principal focus

parallel light and heat rays

convex lens

paper

Figure 6 A convex lens will focus the Sun's rays at one point.

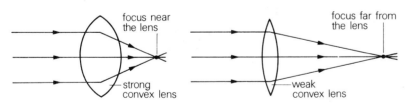

focus near the lens

strong convex lens

focus far from the lens

weak convex lens

Figure 8 The focus is close to a strong convex lens and far from a weak convex lens.

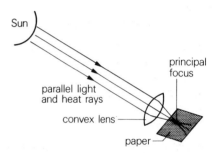

concave lens

parallel beam

rays spread out

Figure 7 Light rays spread out from a concave lens.

Images in lenses

When you look through a lens, the picture you see is called an *image*. Different kinds of lenses make different images:

Concave lenses A concave lens always makes things look smaller, as Figure 9 shows. The image in a concave lens is:
■ always smaller than the object
■ always the right way up
■ virtual – it can never be focused onto a screen.

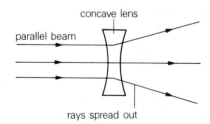

object

small image

concave lens

Figure 9 A concave lens makes an object look smaller.

Convex lenses

Convex lenses give two different kinds of image. When you look at an object *close up*, as in Figure 10, the object looks bigger. The image that you see is:

- magnified
- the right way up
- virtual.

But with an object *a long way off* a convex lens will make a picture or image of it on a screen. This is called a real image. Look at Figure 11. A convex lens is being used to make an image of a distant tree on a screen. The image is:

- smaller than the object
- the wrong way up
- real.

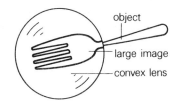

Figure 10 A convex lens makes a close object look bigger.

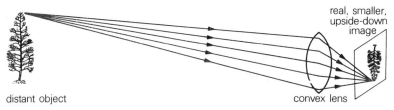

Figure 11 A convex lens will make a real image of a distant object.

So convex lenses can be used to give at least two different kinds of image: a large image of an object close to the lens; a small, upside-down image of an object further away. In the next unit you will see how lenses can be very useful for seeing things in a different light.

Total internal reflection

Sometimes a ray of light travelling in glass is bent so much that it cannot escape into the air. The ray is reflected back into the glass. This is called *total internal reflection*. All of the light is reflected back into the glass.

Underwater The same thing can happen to a beam of light in water. Figure 12 shows light from a special underwater torch meeting the surface of the water. At first the torch beam leaves the water and shines out into the air. But when the beam meets the surface at a smaller angle it cannot escape – the beam is totally internally reflected by the water surface. The water surface is like a perfect mirror.

Total internal reflection only happens when a light ray is trying to leave a 'slower' material and enter a 'fast' one, where it can travel more quickly; for example: from glass to air, or from water to air.

Using total internal reflection

Special reflectors Glass can be shaped into a *prism*. Figure 13 shows a special prism with angles measuring 45°, 45°, and 90°. This type of prism can be used just like a mirror to reflect light rays. Light goes straight through one edge, but when it meets the second edge it cannot escape. The light is totally internally reflected. Two of these prisms can be used to make a periscope. In fact, they are better reflectors than ordinary mirrors.

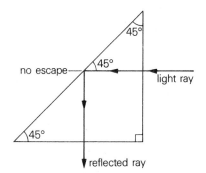

Convex lenses are used in microscopes to give a magnified image. This photo shows two human nerve cells magnified 2000 times.

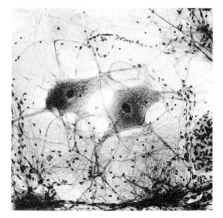

Figure 12 A beam of light can be totally reflected by the water surface.

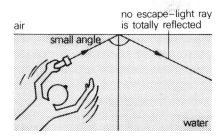

Figure 13 A prism reflects light rays just like a mirror does.

98

These prisms can also be used to make light rays turn round and go back where they came from! Figure 14 shows a beam of light being reflected twice inside a prism, and changing its direction completely. Bicycle reflectors, cat's eyes, and bright diamonds use total internal reflection just like this.

Carrying light Light rays can be trapped inside a solid glass pipe by total internal reflection and so travel around corners. Figure 15 shows a simple 'light pipe' carrying light along a curved path. Light rays cannot escape from the sides of the pipe – they are totally reflected back in. Light carriers like these are called *optical fibres*. Hundreds of optical fibres joined together are used by doctors to see inside people's bodies.

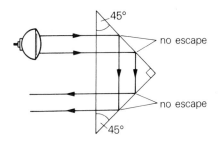
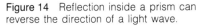

Figure 14 Reflection inside a prism can reverse the direction of a light wave.

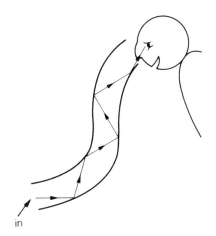

Figure 15 Light can be trapped by total internal reflection. In this way it can travel around corners.

Summary

1 Light rays bend when they travel from one material to another, unless they meet the surface at right-angles.
This bending is called refraction.

2 Refraction in water makes the bottom of a pool look closer than it really is, and a stick in water look bent.

3 Glass can be made into lenses which refract light in special ways:
A *concave* lens is a diverging lens (like a *convex* mirror).
A *convex* lens is a converging lens (like a *concave* mirror).

4 Concave lenses always make a smaller image.
Convex lenses sometimes make a large, upright image – sometimes they make a small, upside-down image.

5 Rays of light in glass or water are totally internally reflected if they meet the surface at a small angle.

6 Total internal reflection is used in special reflecting prisms and in light carriers or pipes called optical fibres.

Exercises

1 Draw and complete these diagrams to show what happens to the beam of light as it enters the glass block, and then leaves it.

2 Give three examples of materials that refract light rays. What happens to the speed of light rays when they enter these materials?

3 Explain why:
a) a stick looks bent when it is lowered into water
b) a swimming pool looks shallower than it really is
c) spear-fishing is difficult.

4 What types of image can a convex lens make? What type of image is always made by a concave lens?

5 Copy out and fill in the blanks:
Convex lenses are _____ in the middle than at the edges. They can turn a parallel beam of light into a _____ beam. Concave lenses are _____ in the middle than at the edges. They turn a parallel beam into a _____ beam. Things always look _____ through a concave lens. A convex lens can be used to make a _____ image.

6 Draw and complete these diagrams to show what happens to the light rays. What is this kind of reflection called? Explain how these 45° prisms can be useful.

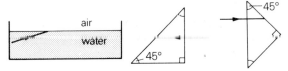

7 How does an optical fibre carry light around corners? Describe how optical fibres can be used by doctors.

4.5 Making light work

Small things can appear larger, distant things appear closer, and pictures can be projected onto a screen. Lenses, telescopes, cameras, and the human eye change the way that light rays travel.

This unit tells you about ways of using light or 'making light work'.

Magnifying glasses

If you hold an object very close to a convex lens and look through it, the object appears much bigger. Figure 1 shows a convex lens being used as a *magnifying glass.* Two rays of light are shown travelling from the top of the pencil and then being bent or refracted by the lens. The person looking through the lens thinks that these rays have come from the larger image – this is how he or she sees the pencil.

In bright light, the pupil of your eye gets very small.

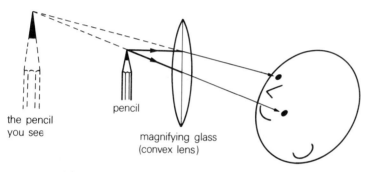

the pencil you see

pencil

magnifying glass (convex lens)

Figure 1   A convex lens gives a magnified image of a close object.

You can find the position and size of the image by drawing the two light rays back without refracting them until they meet. The point where they meet is the top of the image – this is where the top of the pencil seems to be. This kind of diagram is called a *ray diagram*. Ray diagrams are often used in physics to find out the size and position of different images.

Simple magnifying glasses are used by watch makers and menders, stamp-collectors, biologists looking at insects, detectives looking for clues . . . in fact by anyone who wants to make a small object appear larger.

Human eyes

Your eyes use the same type of lens as a magnifying glass, but in a very different way. A magnifying glass makes a large, upright image. Your eye makes a small, upside-down image, as Figure 2 shows.

Focusing   Figure 2 is a very simple diagram showing how the eye lens focuses light rays onto the back of the eye, called the *retina*. The retina is very sensitive to light – it sends messages along the *optic nerve* to the brain. Although the lens always makes upside-down images on the retina, somehow your brain makes things look the right way up!

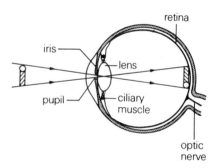

Figure 2   Your eye lens makes an upside-down image on the retina.

This stamp collector is using a simple magnifying glass.

Looking near and far Each eye has special muscles, called *ciliary muscles*, which can make the lens fatter, or thinner. When you look at objects a long way off your ciliary muscles relax and the lens becomes thinner – your eye is not working quite so hard. But light rays from objects very close up need a much thicker lens to focus them onto the retina. The eye muscles must tighten to make the lens thicker, as Figure 3 shows. This is why your eyes get tired if you do a lot of close work.

Night and day Light rays enter your eye through a small opening called the *pupil*. In a dark room, or at night time, your pupil opens wide to let as much light in as possible. In very bright light, like strong sunlight, the pupil gets very small (see photo). Your eye lens can focus more easily on things when the pupil is small – this is why it is better to read in strong light. The size of your pupil is controlled by a coloured muscle called the *iris*. It is the iris that makes your eyes blue, brown, hazel, green, or any other colour.

Helping the eye lens

Short sight Most people can see objects in the distance quite clearly. But some people find that objects a long way off appear blurred and unclear. They are *short-sighted*. Their eyeball is too long. The lens in their eye is focusing light rays *before* they reach the retina, as Figure 4A shows.

Helping short sight Short sight can be helped by spectacles with a concave lens, as Figure 4B shows. The concave lens spreads out or diverges the light rays before they reach the eye. The eye lens then focuses them exactly onto the retina. With spectacles containing concave lenses of just the right strength a short-sighted person can see distant objects quite clearly.

Long sight Most people can see near objects, like this page, quite clearly. But long-sighted people cannot focus properly on near objects. Their eyeball is too short. Light rays from a near object are not focused onto the retina, as Figure 5A shows.

Helping long sight Long sight can be helped by spectacles with a convex lens, as Figure 5B shows. This lens brings in or converges the light rays before they meet the eye. The eye lens then focuses them exactly onto the retina, and close objects can be seen clearly.

Lens cameras

The human eye uses a convex lens to make an image on the retina. Lens cameras use a special glass lens to make an image on a film at the back of the camera, as Figure 6 shows.

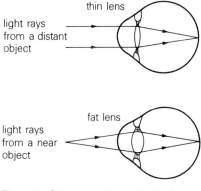

Figure 3 Ciliary muscles make the lens thin to focus a distant object and fat to focus a near object.

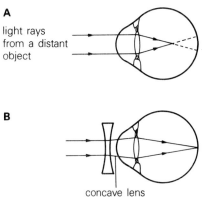

Figure 4 Short sight can be corrected by a concave lens.

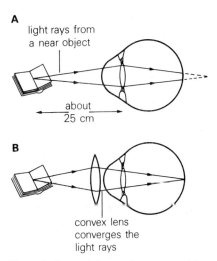

Figure 5 Long sight can be corrected by a convex lens.

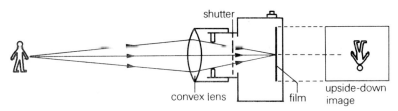

Figure 6 The lens in a lens camera makes an upside-down image on the film.

Most of the time your eyes are open and light is falling onto the retina. But with a camera the opening which lets light into the camera and onto the film is usually closed by a *shutter* (a bit like your eyelid). When someone takes a picture the shutter opens for a split second, just long enough to let some light reach the film, and then it shuts again. To photograph fast-moving objects the shutter must only be open for a very small part of a second, sometimes less than one-hundredth of a second.

Just like the pupil in the eye a camera has a hole to let light in, called the *aperture* (Figure 7). In dim light the aperture needs to be as large as possible. In bright light the aperture needs to be much smaller, otherwise too much light reaches the film.

The eye can focus light rays from near and far objects by making its lens fatter and thinner. Cameras cannot alter the size and strength of their lens — but many cameras can focus light rays clearly by moving their lens backwards (for far objects) and forwards (for near ones).

The human eye and the camera have a lot in common: they both have a lens; the eye has a retina to detect light, the camera has a film; the eye has an opening called the pupil to let light in, the camera's opening is called an aperture. The eye and the camera are both *optical instruments*.

Some other optical instruments

Optical instruments use mirrors or lenses (or both) to change the paths of light rays, by reflecting them or bending them. They can help people to see things more clearly, to make things look larger, to make distant objects look closer, or to focus images onto a screen. Here are a few more examples:

Projectors Projectors use a mirror and two convex lenses to shine a bright beam of light through a film. After the light has gone through the film it passes through another lens that 'projects' a magnified image onto a screen.

Microscopes A microscope uses two strong convex lenses to make very small objects appear much larger. Figure 8 shows a simple drawing of a microscope. It can magnify things much more than a simple magnifying glass. Powerful microscopes can make things appear several thousand times larger than they really are.

Telescopes Most telescopes use two lenses to make distant objects look closer. Figure 9 is a simple drawing of a telescope that uses two convex lenses to look at distant stars and planets. This is called an *astronomical telescope*. It makes things appear upside-down. In 1609 Galileo made a telescope that used one convex lens and one concave lens that made an image the right way up. With it he was the first person to see the craters of the moon, and the satellites of Jupiter.

Modern telescopes use huge concave mirrors to collect as much light as possible from stars and planets, and reflect it through a lens. These are called *reflecting telescopes*, and were first suggested by Sir Isaac Newton in 1704.

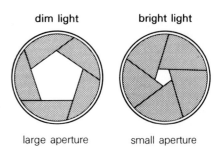

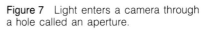

Figure 7 Light enters a camera through a hole called an aperture.

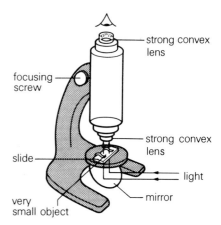

Figure 8 A microscope uses two strong convex lenses to magnify very small objects.

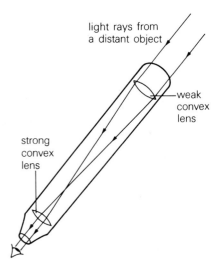

Figure 9 An astronomical telescope uses two convex lenses to make distant objects look closer.

Using other waves for 'seeing'

All optical instruments use light waves for seeing things. Optical telescopes use light rays to see stars and planets. But there are other ways of finding out about outer space:

Radio telescopes Stars give out radio waves as well as light waves. By collecting these waves from distant stars astronomers can learn more about them. Radio waves are collected by enormous radio telescopes that use a giant mirror, just like reflecting telescopes.

Infra-red waves Astronomers also study the infra-red waves from distant stars to find out more about them. This means that there are now three branches of astronomy: optical, radio, and infra-red astronomy.

Infra-red rays can also be used to take pictures in the dark. Ordinary film is sensitive to light rays – but the film used in infra-red photography is sensitive to heat rays. It can be used to take pictures of animals in the dark, an aircraft landing at night, even the heat escaping from a house or a person's head.

This radio telescope is in Parkes, Australia. Its giant mirror collects radio waves from distant stars.

Summary

1 A magnifying glass uses a convex lens to make small things look larger – it produces a magnified, upright image.

2 Human eyes use a convex lens to focus an image onto the retina, at the back of the eye. This lens becomes fatter (stronger) when people look at things close up, and thinner (weaker) when they look at things a long way off.

3 Short-sighted people cannot see distant objects clearly – they need a concave lens to help their own eye lens. Long-sighted people cannot see near objects clearly – they need a convex lens to help focus the light rays.

4 Like the eye, lens cameras use a convex lens to focus light rays. They also have an opening to let light in, called the aperture, which needs to be small in bright light and large in dim light.

5 Projectors, microscopes, and telescopes are other optical instruments which help people to see things in different ways by bending and reflecting light rays.

6 Other waves can be used to give information about objects. Infra-red and radio waves are used in astronomy.

This photo was taken with heat-sensitive film. The heat shadows in the top line of planes (third and fifth from the left) are invisible to the eye. They show that two planes have already left the tarmac. A spy could take an infra-red photograph to work out how many enemy planes took off.

Exercises

1 Draw a simple diagram of the eye and label: the lens, the retina, the iris, the pupil, the optic nerve, and the ciliary muscles.

2 Copy out and fill in the blanks:
A person is short-sighted if her eyeball is too _____ . Spectacles with a _____ lens are needed. A person is long-sighted if his eyeball is too _____ . Spectacles with a _____ lens are needed. These focus light rays exactly onto the _____ .

3 What kind of lens is used for a magnifying glass? Write down three of its uses.

4 What job does the pupil of the eye do?

5 Explain how a lens camera is like the human eye.

6 Write down the names of three optical instruments. Explain simply what each one is used for.

7 What are the three kinds of waves that astronomers use to find out about stars and outer space?

4.6 The colours of the rainbow

Sunlight is really a mixture of many different colours, the colours of the rainbow. Most people think they can see seven. Whenever you see coloured objects – traffic lights, grass, television, or the sky – you are seeing one, or more, of these colours.

Splitting white light

In 1666 Isaac Newton first showed that white light could be separated into different colours when it travels through glass. He used a piece of glass in the shape of a triangle called a *prism*, as Figure 1 shows.

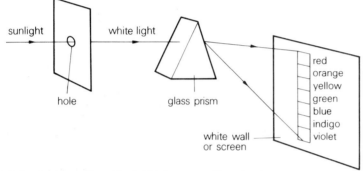

Figure 1   A prism splits white light into seven different colours.

Newton used a glass prism to separate light into its different colours.

The prism splits sunlight into six or seven different colours, that can be seen on a white screen. Newton realised that white light is really a mixture of these different colours.

Every time you see a rainbow you are seeing sunlight split into these separate colours. Two things are needed to make a rainbow: strong sunshine and raindrops in the air. Each raindrop acts like a tiny prism. It splits the white sunlight into the colours of the rainbow. Hundreds of raindrops in the sky act together to form the curved 'bow' that you see.

The same colours are seen when sunlight travels through the water from a hose pipe or the drops from a tap.

Different speeds   All the colours that make up white light travel at about the same speed in air. When they go into glass or water they are slowed down, and so they are bent or refracted. But each colour is slowed down by a *different amount*. Violet is slowed down the most, so it is bent the most. Red light is not slowed down as much, so it is not bent as much by glass or water. Figure 2 shows how red and violet light are bent or refracted by a glass prism and a raindrop.

The visible spectrum   You saw in unit 4.1 that light belongs to a family of waves called the electromagnetic spectrum. Light waves make up a small part of that family – this part is called the *visible spectrum*. It is shown in Figure 3.

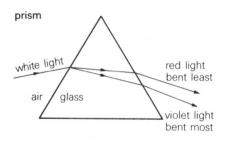

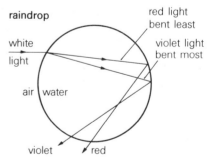

Figure 2   Red light travels faster than the other colours of light through a prism or through water. This means that red light is not refracted as much as the other colours. Violet light is refracted the most.

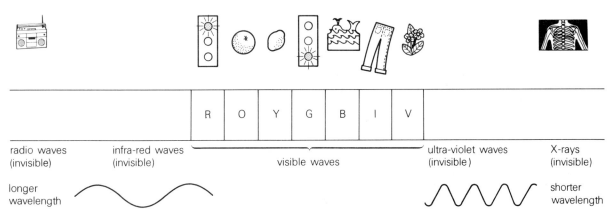

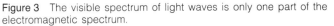

Figure 3 The visible spectrum of light waves is only one part of the electromagnetic spectrum.

The visible spectrum is made up of seven different colours: red, orange, yellow, green, blue, indigo, and violet. (You can remember them by: 'Richard Of York Gave Battle In Vain'.) Each one of these colours travels at a different speed in glass – this is why a glass prism separates them. Each colour also has a different *wavelength*. Red light has the longest wavelength, violet light the shortest.

Finally, the colours inside the visible spectrum are the only waves that can be seen by the human eye. On one side of the spectrum there are infra-red rays – you can feel these but your eye cannot see them. On the other side there are ultra-violet rays. These rays give your skin a sun tan, but they are invisible.

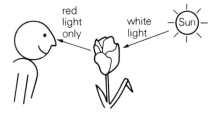

Figure 4 A red flower reflects red light and absorbs all the other colours.

Seeing colours

Ordinary white light, like sunlight, is really a *mixture* of all the colours of the rainbow. So when you look at a red flower in sunlight, why does it look *red*? As Figure 4 shows, the only light reaching your eye from the flower is red light. The rest of the colours mixed together in sunlight have been absorbed or 'swallowed up' by the red part of the flower. Only the red light in the spectrum is reflected from it. A daffodil absorbs all the colours in white light except yellow – grass absorbs them all, except for green. What about white paper, or snow?

Black and white objects White objects do not absorb any of the colours in the spectrum. White light is reflected totally. Black things are the opposite. Coal, soot, or black paper absorb *all* the colours in the spectrum so that none of them can reach your eyes.

Seeing things in a different light White paper reflects every colour, so it looks white in sunlight. But how would it look under a yellow light, like the light from a street lamp? Only yellow light is reaching the white paper, so only yellow light can be reflected from it. This means that white paper looks yellow under a yellow light, as Figure 5 shows.

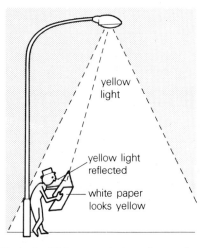

Figure 5 White paper looks yellow under a yellow light.

What colour would a blue jumper look in different coloured lights? Being blue it absorbs every colour of the spectrum *except* blue light. So if you see it in white light the jumper looks blue. Suppose you see a blue jumper in a red light. It will not reflect *any* of the red light – all the light falling on the jumper will be absorbed, as Figure 6 shows. In other words the jumper will look black.

A blue jumper looks black in red light. A red jumper looks black in blue light because none of the blue light is reflected by red. You only see things in their 'true colour' in white light, which contains all the colours of the rainbow.

Mixing colours

Making white   White light can be split into seven different colours. Figure 7 shows how these separate colours can be put back together again. When a wheel with the seven colours on it spins quickly enough it appears to be white. Your eye puts the colours back together again. This spinning wheel is called Newton's disc.

In fact, only three colours are needed to make you see white light: red, green, and blue. These are called the three *primary colours*. If red, green, and blue lights are shone onto a white screen the lights mix together to produce white light where all three lights meet. Figure 8 shows the colours made where only two of the lights overlap. (Note: you only get this result with coloured lights, not paints.)

Watching colour television   These three primary colours can be used to make almost any colour, when they are mixed together in the right amounts. The screen of a colour TV is covered by thousands of small red, green, and blue strips. Your eyes collect the light from these glowing strips and mix it to give you the sensation of yellow, green, pink, turquoise, and so on.

One nineteenth-century artist, called George Seurat, did all his paintings using only red, green, and blue dots. The eyes of the person looking at the painting did Seurat's mixing for him!

What is light?

Here are some true sentences about light:
■ Light is a form of energy.
■ Light is a wave that travels through space.
■ Light is part of the electromagnetic spectrum.
■ Light is what you see by.
■ Light travels in straight lines but it can go around corners.
■ Light is a stream of moving particles.

Scientists know a lot about how light behaves, how fast it travels, and how it can be used, but they are still not sure what it *is*. Newton believed that light was a fast-moving stream of little bits called 'corpuscles'. Later scientists, in the 19th century, said: 'No, Newton was wrong, light is a wave.' Then, in 1905, Albert Einstein claimed that Newton was right after all: 'Light is a stream of moving particles.' Today scientists believe that light behaves sometimes as a wave, sometimes as a particle.

Figure 6   A blue jumper looks blue in white light but black in any other coloured light.

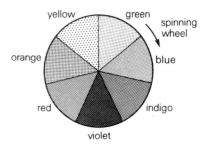

Figure 7   If you watched Newton's disc spinning, you would see white light.

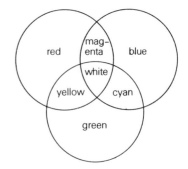

Figure 8   Red, green, and blue are the primary colours. When they are all mixed together they make white light. Look at the colour photo on the back cover.

Summary

1 White light can be split into separate colours by a prism or a raindrop.

2 The colours separate because they travel at different speeds in glass or water.

3 Coloured objects reflect some of these colours, but absorb others. White objects reflect all the colours – black objects reflect none.

4 The separate colours can be mixed together again by spinning Newton's disc, or by overlapping red, green, and blue light beams.

5 Who knows what light really *is*?

Einstein supported Newton's theory that light consists of particles.

Exercises

1 Copy out and fill in the blanks:
Isaac _____ first split sunlight into different _____ , using a _____ . In a rainbow, every _____ acts like a tiny _____ . The seven colours of the spectrum are _____ .

2 Explain why:
a) a red flower looks red
b) daffodils look yellow
c) grass appears green
d) black objects look black
e) white paper looks yellow under a street lamp.

3 a) Which colour is refracted most by a glass prism?
b) Which colour is refracted least?
c) Explain why the colours are separated, using a diagram.

4 What are the three primary colours? What colour do they make when added together?

5 Explain what Newton's disc is and how it works.

6 How is a colour television picture made up?

7 a) What is meant by the visible spectrum?
b) What family of waves does the visible spectrum belong to?
c) Which two waves (or rays) in this family lie on each side of the visible spectrum?

8 a) Which colour of the spectrum has the longest wavelength?
b) Which colour has the shortest wavelength?

9 The photo below shows part of a painting by George Seurat. Describe how Seurat did his paintings.

4.7 Sound waves

How does a bat see in the dark? Why do cinemas and theatres have padded walls? How do ships find the depth of water below them? The answers all depend upon another type of energy-carrier: the sound wave.

Hearing sounds

Vibrations   Whenever something moves backwards and forwards, again and again, it *vibrates*. If you 'twang' a ruler over the edge of a table, as shown in Figure 1, it moves quickly up and down. It is vibrating. If you push the ruler further over the edge it vibrates more slowly – the longer the ruler, the slower the vibrations. Each time you twang a ruler you are making a *sound wave*. Sound waves are always made by vibrations. Some more examples include: a drumskin, a guitar string, and a tuning fork. You cannot see the ends of a tuning fork vibrating – the vibrations are too fast. But they make quite a splash if they touch the surface of some water.

Hearing things   You can hear all these vibrations because they are carried through the air to your ear as sound waves. Your outer ear collects the sound, as Figure 2 shows, and carries it to your *ear drum*. The ear drum is like a thin drumskin – whenever you hear a noise the ear drumskin starts to vibrate backwards and forwards. This vibration travels through to your inner ear which sends messages to the brain.

Sound energy   You can only hear sounds when your ear drum vibrates. Very weak, quiet sounds cannot be heard because they do not have enough sound energy to move the skin of the ear drum. But very strong, loud sounds can be dangerous – they may even carry enough energy to break the drumskin of the ear.

In short:
- sound waves are made when something vibrates
- they carry energy from one place to another
- they can often be detected by your ears.

How sound waves travel

'Stretches' and 'squashes'   Sound waves always need something to travel in – they cannot travel through empty space. There must be many objects in the Solar System making noises. But you will never hear these sounds because there is nothing to carry them. Figure 3 shows how sound waves are carried through the air from a drum.

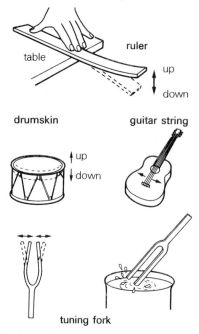

Figure 1   Sound waves are made by vibrations. Vibrations are made when something moves backwards and forwards.

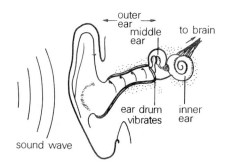

Figure 2   You only hear sounds when your ear drum vibrates.

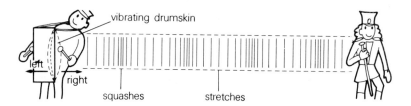
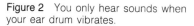

Figure 3   Sound waves travel by 'stretching' and 'squashing' air particles.

The drumskin vibrates backwards and forwards, making the air vibrate. At first it moves to the right, squashing the air next to it. As it vibrates back again, to the left, the air next to the drumskin spreads out. It stretches to fill more space. The air next to the drumskin is squashed and then stretched every time the skin moves backwards and forwards. This air squashes and stretches the air next to it, and so on. These 'squashes' and 'stretches' travel through the air until they reach your ear drum, which starts vibrating itself.

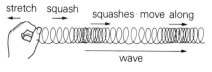

Figure 4 'Squashes' and 'stretches' travel along a slinky spring like a sound wave through air.

Slinky spring   You can see how a sound wave travels by squashing, then stretching, a slinky spring. Figure 4 shows how these stretches and squashes travel along the spring, just like a sound wave travels through the air.

Different materials   Sound waves are travelling vibrations. Most materials can carry these vibrations: air, water, wood, steel, string, brick can all carry sound waves. Solids carry them best of all: try knocking on a wooden table. The sound is much clearer when you place your ear firmly on the table.

Sound travels faster through solids than it does through liquids, and *much* faster in solids than in air. Figure 5 shows some different *speeds of sound*. In air, sound travels about 1 km in 3 seconds. You can use this to guess how far you are from a thunderstorm.

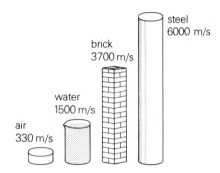

Figure 5 Sound travels at different speeds in different materials. It travels fastest in solids.

Echoes

Just like water waves and light waves, sound waves can be *reflected* when they meet a barrier. Figure 6 shows how you can hear a watch ticking by reflecting the sound off a sheet of wood or glass. The ticking sounds loudest when the two cardboard tubes are held at the same angle to the barrier. The incident angle then equals the reflected angle. Sound waves obey the same law of reflection as light waves.

Good and bad reflectors   Shiny surfaces are the best reflectors of light and heat waves – *hard, solid surfaces* are best for reflecting sound waves. If a girl shouts when she is surrounded by high walls, cliffs, or mountains, she hears an echo. Echoes are just sound waves reflected by a solid surface. Soft surfaces are bad reflectors of sound. Cinemas and theatres have soft, padded walls and thick carpets to absorb sound and stop unwanted echoes.

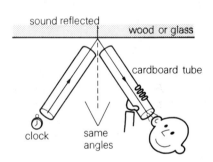

Figure 6 Sound waves can be reflected off a barrier. They obey the law of reflection – the incident angle is the same size as the reflected angle.

Using echoes   Echoes can be a nuisance in a large hall, cinema, or theatre. But they can be useful at sea, as Figure 7 shows. Ships send out sound waves which are reflected by the sea-bed and collected again. The longer the echo takes to come back, the deeper the sea-bed. If the echo comes back very quickly then the sea-bed must be very close. This process is called *echo-sounding*. Submarines also use echo-sounding to 'see' in front of them, by sending out sound waves and picking up the echo. Even fishing boats use echoes – sound waves are reflected by large shoals of fish.

Dolphins and bats send out very high-pitched noises to find out what is in front of them. This is also how bats 'see' – they listen to the echoes reflected back to them.

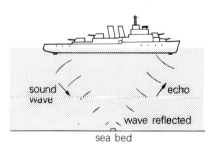

Figure 7 The depth of the sea can be measured by reflecting sound waves off the sea-bed.

Pitch and frequency

Fast and slow vibrations   All sounds come from something
vibrating. If it vibrates backwards and forwards, or up and down,
very *quickly* it makes a *high* pitched sound. Very few people can
hear a bat's squeak – it is too highly pitched. The ruler in
Figure 1 is making a lot of vibrations every second. A longer ruler
goes up and down more slowly. It makes fewer vibrations every
second, and gives a lower-pitched sound: In other words:

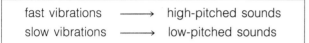

fast vibrations	$\longrightarrow$	high-pitched sounds
slow vibrations	$\longrightarrow$	low-pitched sounds

The pitch of a sound depends upon the number of vibrations every
second.

The limits of hearing   Human beings cannot hear sounds from
objects that are vibrating very, very quickly, although dogs, bats,
and dolphins can. Most adults can hear sounds as low as 25
vibrations per second, and as high as 18 000 vibrations per second.
Young children can hear sounds even lower, and higher, than
these. Dogs can hear whistles pitched higher than 20 000 vibrations
per second.

Frequency   Sounds above the limits of human hearing are called
ultrasonic. They vibrate a large number of times every second –
they have a very high *frequency*. The pitch of a sound depends
upon its frequency – high-pitched notes have a high frequency,
low-pitched notes have a low frequency.

Frequency is measured by the number of vibrations every second
with a unit called *hertz* (Hz for short):

e.g.   a low note: 50 hertz   =   50 vibrations per second
      a high note: 10 000 Hz   = 10 000 vibrations per second
ultrasonic note: 25 000 Hz   = 25 000 vibrations per second

Notes of different frequencies can be arranged in a special order to
make a musical *scale*.

The sound of music

Octaves   Notes of different frequencies are arranged on a piano
keyboard as shown below. The gap between two notes is an *octave*
if one has got twice the frequency of the other:

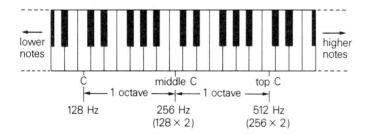

Musical instruments   All musical notes are made when something
vibrates. Pianos use vibrating strings to make their musical notes.

In a pipe organ, different sized pipes give
notes of different pitch.

A bat navigates by making high-pitched
noises which bounce off hard surfaces as
echoes.

Soft padded surfaces, like this ceiling in a
concert hall, absorb unwanted echoes.

Some instruments use pipes full of air – the air inside the pipes vibrates. A drum uses a vibrating skin, a bell has sides that vibrate, a xylophone uses vibrating bars, a guitar has vibrating strings. The strings or pipes can be adjusted to make notes of different frequencies, as shown in Figure 8.

Figure 9 shows how you can produce high and low notes by making the air in a milk bottle vibrate.

Using your voice The human voice uses two things to make a musical note: folds of 'skin' called *vocal cords* which vibrate back and forth; and a short pipe called the *voice-box*, or 'Adam's apple', which helps to make the sound louder. When you make a high-pitched sound, like a scream, your vocal cords are stretched very tightly, and they vibrate quickly. With lower notes the vocal cords are much slacker, and the vibrations slower. Men have longer and thicker vocal cords than boys and women. This means their voices can make deeper, low-pitched notes.

Your own voice can make both high and low frequency sounds.

	strings	pipes
low notes	long string	long pipe
	slack string	
	thick string	
high notes	short string	short pipe
	taut string	
	thin string	

Figure 8 The strings or pipes of a musical instrument can be adjusted to make notes of different frequency.

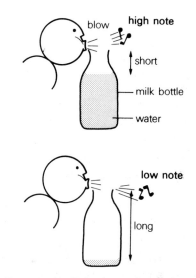

Figure 9 A milk bottle can be used as a musical instrument. By filling eight milk bottles with different amounts of water, and arranging the bottles in order, you can make a simple organ.

Summary

1 Sound waves are made by vibrations.

2 Sound waves move through different materials by 'stretching' and 'squashing' them as they travel along. Sound always needs a material to travel in, but it travels at different speeds in different materials.

3 Reflected sound waves are called echoes. Echoes are used by ships, submarines, bats, and dolphins.

4 Slow vibrations make low-pitched sounds, fast vibrations make high-pitched sounds. The pitch of a note depends upon its frequency, measured in hertz.

5 All musical instruments use vibrations to make notes of different frequency. The longer, thicker, and slacker a string the lower the frequency of its note.

Exercises

1 Copy out and fill in the blanks:
Sound waves are always made by _____ . Two examples are _____ and _____ . Your ear _____ vibrates when a sound wave reaches it. Your ear can be damaged by a wave which carries a lot of _____ .

2 Write down the speeds of sound in:
a) air   b) water   c) steel
How far would sound travel in 10 seconds in each of these materials?

3 a) What is an echo?
b) What kinds of surfaces are best for making echoes?
c) What kinds of surfaces absorb sound waves and stop echoes?

4 Explain how sound waves travel through the air.

5 a) When are echoes a nuisance?
b) Explain how echoes can be useful at sea.

6 Some musical notes have a frequency of 128 hertz. What does this mean? What is the frequency of a note one octave above this?

7 Write down examples of musical instruments that use:
a) air vibrating inside a pipe
b) a vibrating string
c) a vibrating skin.
Describe how strings can be changed to produce high notes and low notes.

8 Explain how your vocal cords make sounds, and how the frequency can change.

Lasers:

Life rays or death rays?

In 1960 an American scientist, Theodore H. Maiman, demonstrated the world's first *laser*. Since then laser beams have found many uses, some peaceful, some not so peaceful. But first of all . . .

What is a laser?

A laser is an instrument that makes a very concentrated beam of *light*. Some laser beams have such power they can burn holes in steel plates and set charcoal on fire. Yet a laser beam can be controlled so precisely that surgeons use it to operate on the back of the human eye. A laser beam is so straight, and so narrow, that if it were sent from the Earth to the moon (384400 km) the spot on the moon would only be about 2 km wide.

All the light waves in a laser beam are exactly the same colour and they all have exactly the same wavelength.

Making a laser beam

The main part of T. H. Maiman's first laser was a large ruby. This large ruby is set between two mirrors:

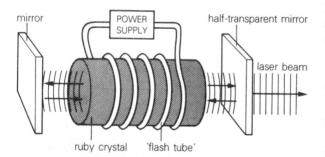

mirror

POWER SUPPLY

half-transparent mirror

laser beam

ruby crystal   'flash tube'

A long 'flash tube' is wrapped round and round the ruby. This flash tube produces a very strong light. The ruby *absorbs* this light energy. Then the light beam passes backwards and forwards through the ruby, bouncing off the mirrors at each end. As it travels to and fro the beam picks up more and more energy from the ruby. The ruby is a kind of *energiser*. The beam gets stronger and stronger, redder and redder.

One of the mirrors is half-transparent. It lets some light through. Out of this mirror comes the *laser beam*. A ruby laser makes a bright red laser beam. Many other kinds of laser have been

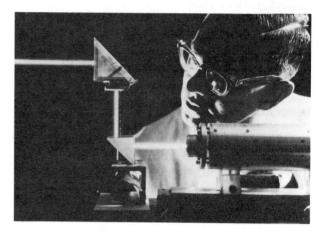

The laser beam from this ruby laser is being reflected through two prisms.

made since 1960. Some use a solid (like ruby), some use a tube of gas, like helium–neon, argon, or carbon dioxide. They all have their own colour and wavelength. But all lasers have three important parts:
- a *source* of light, like the flash tube
- an *energiser*, solid or gas, to turn ordinary light into a strong laser beam
- two *mirrors* to send the light to and fro through the energiser.

By the way, do you know what the word **LASER** stands for?: "**L**ight **A**mplification by **S**timulated **E**mission of **R**adiation" – quite a mouthful!

Travelling light

Light beams and laser beams can be carried round corners in special tubes called *optical fibres*. Several optical fibres are joined together to make a *fibre optic bundle*. This bundle is very narrow – it can be pushed inside the body or a machine (see photo on page 113). Doctors can now look inside a stomach, a knee joint, or a woman's womb.

Optical fibres thinner than a human hair can carry a laser beam deep inside the human body. The laser can be directed onto one tiny spot. In this way cancer cells can be destroyed, leaving their healthy neighbours unharmed.

Most scientific inventions have peaceful uses and wartime uses. Unfortunately most of the money needed to develop the inventions seems to go towards wartime uses. The following table shows some peaceful and wartime uses of lasers.

The laser beam: friend or foe?

'FRIENDLY' USES

Treating cancer   Lasers are being used to cure certain types of skin cancer. The cancerous growth absorbs the laser beam better than the healthy skin around it. The heat energy of the laser beam kills the dangerous growth. Lasers can even be used to kill warts!

Eye surgery   The retina at the back of the eye sometimes gets damaged and broken. A laser beam can go safely through the front of the eye and be focused onto the retina. It makes a tiny spot less than 1 mm wide. The heat from this spot joins, or welds, the retina together again, and makes a tiny scar.

At the dentist   Laser beams have been used to kill tooth decay in a fraction of a second. The dark, decayed parts of a tooth absorb the laser beam more than the white, healthy parts. And it doesn't hurt!

Carrying messages   A laser beam can be used to carry messages quickly and accurately from one part of the Earth to another. It can 'carry' sound waves, radio waves, and TV waves and can be bounced off a satellite.

Cutting diamonds   The beam from a carbon dioxide laser can cut steel and even drill holes in diamonds. This means that lasers can be used in hundreds of different industries.

Photos in 3-D   Photographs in 3-dimensions can be taken by a laser beam. They are called *holograms.*

NOT SO FRIENDLY USES

Laser beams can be very dangerous. Millions of pounds have been spent in developing the laser as a weapon. It can have many uses in warfare. Here are just a few . . .

Death rays   By 1965, America had already spent more than 30 million dollars on making a laser into a weapon. The beam from a carbon dioxide laser can set a soldier's clothes on fire from a few miles away.

Destroying planes and missiles   Very recently a powerful laser has been tested which can shoot down an enemy plane. Some people believe that a laser could be used to shoot down an enemy missile. It would need to be very powerful – perhaps millions of watts. The Armed Forces are spending a lot of money trying to make an anti-missile laser.

Finding a target   A laser can be beamed at an enemy 'stronghold'. The reflections are picked up by an aircraft. The beam's reflection can pinpoint the exact position of the enemy.

Range finders   Echoes are used to find the depth of the sea. 'Laser echoes' can be used in the same way for finding the exact position of a target. They can tell soldiers *exactly* how far away a target is. Laser beams can also be used as a kind of 'radar'. A beam is sent out and bounces back off a solid object. The laser echo can be used as a warning.

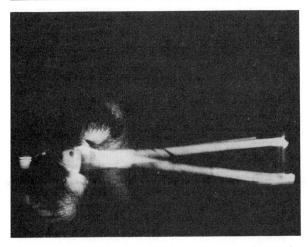

Laser beams can be fired into the eye to mend a damaged retina.

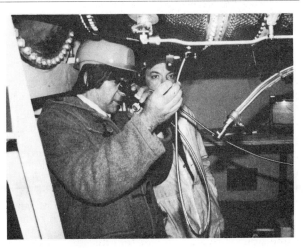

This man is inspecting the inside of a gas turbine by looking through an optical fibre.

Topic 4 Exercises

More questions on waves

1 This diagram shows waves in a ripple tank about to meet three different kinds of barrier. Draw diagrams to show what happens to each wave.

2 a) A girl with long sight cannot focus light rays from a near object onto her retina. Draw a simple diagram to show this. Draw another diagram to show how this girl can be helped with a lens.

b) A boy with short sight cannot focus light rays from a distant object onto his retina. Draw a simple diagram to show this, and another diagram to show how the fault can be corrected with a lens.

3 This diagram shows a ray of white light about to meet a prism. Draw, complete, and label the diagram to show what happens next.

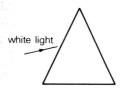

white light

4 Suppose three circles of coloured light shine onto a white screen, as the diagram shows. What colours will be seen at A, B, C, and D?

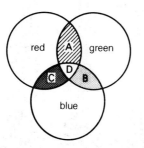

red   A   green
    C D B
       blue

5 A ship on the sea sends out a sound wave. After 1 second the echo comes back. If the speed of sound in sea water is 1500 m/s, how deep is the sea? A few minutes later suppose the echo comes back after only $\frac{1}{2}$ second. How deep is the sea there?

Things to do

1 **The disappearing coin**

You need: a coin, a jam jar, water.
Place the coin under the jar. Look through the *side* of the jar and you can see the coin. Now almost fill the jar with water. The coin disappears.
Next, wet the coin itself with water. The coin appears again!

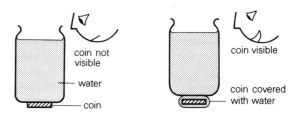

coin not visible

water

coin

coin visible

coin covered with water

This trick depends upon the refraction of light rays by water. Another trick using refraction is described on page 96.

2 **Making your own lens**

You need: a nail, some cooking foil, water.
Make a tiny hole in the foil with the point of the nail. Let one drop of water fall gently onto the hole so that it stays there. The drop acts as a tiny but strong lens. You can use it as a magnifying glass to look at small objects.

3 **Making music**

a) **A bottle organ**
Figure 9 on page 111 describes how to make a simple organ from milk bottles.

b) **Drinking straws**
A paper drinking straw can be used to give notes of different frequency. Flatten the end of the straw and thin its corners off with scissors. Blow gently through this end. Shorter straws give higher notes.

4 **Making your own 'telephone'**

You need: two tins, a long piece of string. Punch a hole in the bottom of each tin and thread the string through. Tie knots inside the tin. Stretch the string tightly and get someone to talk through the tin at one end. You can hear the voice quite clearly. The string carries the vibrations to your ear.

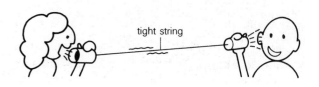

tight string

Atoms and electrons

Electricity is carried to homes and factories by huge power pylons.

5.1 The atom

Until this century no-one knew much about atoms. They were thought to be the smallest, indivisible particles of matter. But in the last 80 years scientists have discovered a lot about what goes on 'inside' the atom, as this unit explains.

What is an atom?

Everything around you is made up of atoms – the air you breathe, the pages of this book, your own body all contain millions and millions of tiny particles called atoms. The atom is the 'building brick' of all the materials around you.

Atoms and molecules   Every word on this page is made by using one or more of the 26 letters in the alphabet. Most words use more than one letter (like 'the'), some use only one letter (like 'a'), some use letters more than once (like 'eye'). In the same way every material on Earth is made by using one or more of the 92 atoms that occur naturally. Most words use *different* letters joined together – most materials are made up of different atoms joined together. Atoms joined together are called *molecules*. Remember: atoms are like letters, molecules like words.

Naming atoms   Every atom has its own name, number, and symbol. Figure 1 shows some members of the 'atomic alphabet'. Number 1 is hydrogen, H for short; number 8 is oxygen, O for short.

Some atoms do not like going around on their own – they usually join together. Two atoms of hydrogen join together to make one molecule of hydrogen. Two atoms of hydrogen join onto one atom of oxygen to make a water molecule, H_2O for short. Figure 2 shows how a water *molecule* comes from hydrogen and oxygen *atoms*.

Pictures of the atom

Atoms are so small that nobody has ever seen one, even with the most powerful microscope. One hundred million atoms lined up side by side would not stretch as far as this line: ———. So how can scientists tell what atoms are like? They have had to build up a picture of the atom, partly by guesswork and partly by using clues from certain experiments.

The billiard ball picture   One of the ancient Greeks, Democritus, first suggested the idea of the atom in the year 460 BC. If you cut a piece of iron up into smaller and smaller pieces, Democritus said, eventually you end up with pieces so tiny they cannot be cut any more. These uncuttable or 'unsplittable' pieces Democritus called *atoms*. Nobody took him very seriously until 1808 when an Englishman, John Dalton, said that Democritus was right. He also suggested that each material is composed of its own kind of atom: iron contains iron atoms, copper contains copper atoms, and so on. Both Democritus and Dalton thought that atoms were hard, solid, indestructible balls like marbles or billiard balls.

name	number	symbol
hydrogen	1	H
helium	2	He
carbon	6	C
oxygen	8	O
uranium	92	U

Figure 1   Every atom has its own name, number, and symbol. There are 92 different atoms found naturally.

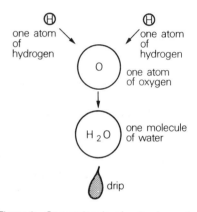

Figure 2   One molecule of water is made up of two hydrogen atoms and one oxygen atom.

John Dalton

Rutherford's picture   In 1911, a man from New Zealand called Ernest Rutherford did experiments which showed that the atom is mostly empty space, as Figure 3 shows. Every atom has a hard, heavy centre called the *nucleus*. Going around and around the nucleus are tiny particles called *electrons*. The simplest atom of all, hydrogen, has one electron orbiting the nucleus. Helium has two electrons, carbon has six, uranium has 92 electrons in each atom.

Holding atoms together

The Earth and the planets go around the Sun. Gravity pulls the planets around in their orbits. Electrons go around the nucleus. But what force keeps the electrons orbiting the nucleus of an atom?

Static electricity   If you rub a plastic pen or comb on your sleeve it can sometimes be used to pick up small pieces of paper (Figure 4). The paper is attracted to the comb by an *electrostatic force*. The same kind of force holds electrons in their orbit around the nucleus. The nucleus has a *positive* charge (+), electrons have a *negative* charge (−). These two charges attract each other, as Figure 5 shows.

Inside the nucleus   By 1932 scientists knew that the nucleus itself was made from two different particles: the *neutron*, and the *proton*. Neutrons are neutral particles – they do not carry any charge. But every proton carries a positive charge which is why the nucleus has a positive charge. The positive charge on a proton exactly balances the negative charge on an electron. So hydrogen atoms have one proton in their nucleus, helium atoms have two, carbon atoms have six, uranium atoms have 92 protons. The number of protons is always the same as the number of electrons in an atom, as Figure 6 shows. Every atom except the simplest hydrogen atom has some neutrons in its nucleus. Helium has two neutrons in the nucleus of its atom, uranium usually has 146. You will see how important these neutrons are in the next section. But first here are the important rules about this picture of the atom:

- All atoms have a nucleus in the middle with one or more electrons circling around it.
- The nucleus is made up of protons (+) and neutrons (neutral).
- Electrons (−) are held in their orbits by their attraction to protons.
- The number of protons in the nucleus is always the same as the number of electrons going around it.

Atoms normally keep to these rules. When they don't amazing things can happen . . .

Splitting atoms up

Smashing the nucleus   Democritus was wrong. Atoms can be split. In 1939 scientists used moving neutrons to break open the nucleus of large atoms like uranium. Moving neutrons are like high-speed bullets. The nucleus of an atom is held together by a very, very strong force which scientists know little about. They just call it the 'strong nuclear force'. A moving neutron can sometimes split open a uranium nucleus. This splitting is called *nuclear fission*. When the nucleus breaks tremendous energy is released. This energy is called *atomic energy* or *nuclear energy*.

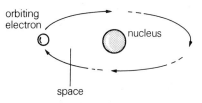

Figure 3   An atom is mostly empty space. The atom with the simplest structure is hydrogen. It has only one electron orbiting the nucleus.

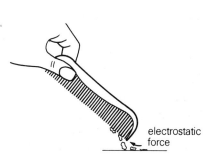

Figure 4   Pieces of paper can be attracted to a plastic comb by an electrostatic force.

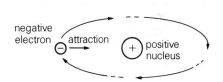

Figure 5   Electrons are held in orbit around the nucleus by an electrostatic force. This force of attraction holds the electrons onto the outside of an atom, like an invisible string would.

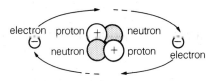

Figure 6   Every atom has the same number of protons as electrons. The helium atom has two protons.

A chain reaction There are two kinds of uranium atom. One kind, U-235, has 143 neutrons in its nucleus. The other kind, U-238, has 146 neutrons. U-235 atoms can be split by a moving neutron, as shown in Figure 7. Each time the nucleus splits it sends out 2 or 3 neutrons. These 3 neutrons might hit three more uranium atoms – they would split and send out 9 neutrons. If there is enough uranium these 9 neutrons will hit nine more atoms, making 27 flying neutrons. This could go on and on, each time releasing more and more energy. With a piece of U-235 any bigger than a tennis ball the chain reaction gets completely out of control, releasing an enormous amount of energy.

Atomic energy The discovery of this chain reaction led to the first *atomic bomb*. It was tried out in the desert of New Mexico, and first used in war against Japan on 6 August 1945. Over 80 000 people were killed. But atomic energy can be used in peace time. Slowing down and controlling the chain reaction can give a steady supply of heat energy. Some scientists hope that this heat energy will save the burning of valuable coal and other fuels.

Rays and flying bits from atoms

Radioactivity Some materials are made up of atoms which send out flying particles or rays. They are called *radioactive materials*. Plutonium, uranium, polonium, and radium are four examples. There are many others. The particles and rays given off are called *radiation*.

Radiation from the nucleus Funny things can happen inside the nucleus of radioactive atoms. They can break up, bits can fly off. These 'flying bits' are of two types called *alpha (α) particles* and *beta (β) particles*.

Alpha particles are made up of two protons attached to two neutrons – in other words a helium nucleus. They can only travel a short distance through the air, and can be stopped by human skin or a thin sheet of paper. Beta particles are fast-moving electrons ejected from the nucleus of atoms. A thin sheet of metal is needed to stop them. (Note: alpha and beta particles are sometimes called 'alpha rays' and 'beta rays'.)

The nucleus of some atoms can send out dangerous rays which travel at the speed of light. They can never be completely stopped, even by thick sheets of lead. These rays are called *gamma (γ) rays*.

Figure 8 shows you the three types of radiation that can come from the *nucleus* of atoms in radioactive materials.

Rays and their uses Another type of ray can be made to come from some atoms. They are called *X-rays*. These are made in special X-ray tubes, when atoms of tungsten are bombarded by electrons.

Gamma rays and X-rays are both part of the electromagnetic spectrum. They both travel at the speed of light, and both have a very short wavelength. Both rays can be very dangerous. Radiation from the first atomic bomb dropped in Japan killed almost as many people as the explosion itself. But both rays can also be useful.

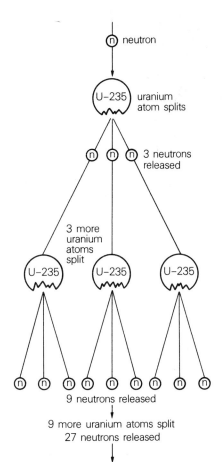

Figure 7 U-235 atoms can be split by moving neutrons which set up an explosive chain reaction.

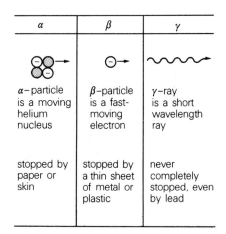

Figure 8 Three types of radiation can come from the nucleus of a radioactive atom. They are alpha particles, beta particles, and gamma rays.

X-rays will travel through human skin and flesh but not so easily through metal. They can be used to take special 'X-ray pictures' of the human body (see photos on pages 3 and 83). Gamma rays are used in some hospitals to kill cells that have been affected by cancer.

Electrons

Electrons are easily set free and released from atoms. This is why electrons were noticed, by a man called J. J. Thomson, years before the atom was split. His discovery of the electron in about 1900 started a completely new branch of physics: *electronics*. It has led to television, radio, amplifiers, and now electronic games and computers.

But scientists had been studying the way that electrons travel, without really knowing it, since 1800. Volta, Faraday, and many others used *electricity* in the 19th century before people knew about atoms and electrons. In the next unit you will see that electricity comes from electrons on the move.

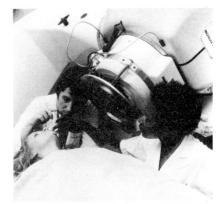

Gamma rays can be used to help people with cancer. This girl is about to have radiotherapy to her neck.

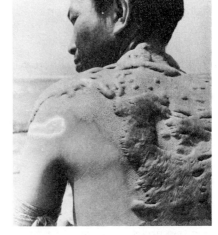

This man survived the atomic bomb dropped on Hiroshima but he suffered terrible injuries. Many people not killed at the time died later from radiation sickness.

Summary

1 Every material around you is made up of atoms. Atoms often join together to make molecules.

2 Every atom has a nucleus of protons and neutrons at its centre with electrons circling around it. This is how to 'picture' the atom.

3 The nucleus has a positive charge which attracts negative electrons. This attraction holds the electrons in the atom.

4 The parts of the nucleus are held together by very strong forces. When this hold is broken the atom splits and terrific energy is released.

5 Three types of radiation can come from the nucleus of certain radioactive atoms — alpha, beta, and gamma rays. Another ray, the X-ray, can be produced by X-ray tubes.

6 Electrons can be taken away from atoms. The electrostatic force holding them onto the atom is weaker than the strong nuclear force.

Exercises

1 What is the difference between an atom and a molecule?

2 Who first suggested the idea of atoms?

3 Explain the difference between the billiard ball picture of the atom and Rutherford's picture.

4 Draw a diagram of a hydrogen atom and label the parts.

5 Copy out and fill in the missing words:
The centre of an atom is called the _____ . It has a _____ charge. Electrons have a _____ charge. Neutrons are _____ . All atoms have the same number of _____ as they do _____ . The three types of radiation from the nucleus are _____ particles, _____ particles, and _____ rays.

6 Describe how atoms can be split. What is a 'chain reaction'?

7 Which type of radiation:
a) travels at the speed of light?
b) contains the same parts as a helium nucleus?
c) consists of fast-moving electrons?
d) will not penetrate human skin?
e) goes through skin, but is stopped by metal?
f) is never completely stopped, even by lead?

8 Explain, in your own words, what gamma rays and X-rays can be used for.

9 Explain the difference between the two kinds of uranium atom.

5.2 Electrons and electricity

Everyone uses electricity. It drives motors and light bulbs, dries hair, and even cooks the Sunday dinner. But what is it? And what has electricity got to do with atoms?

In this unit you will see that when electrons break loose from atoms electricity is made – it can make balloons stick to a ceiling, sparks jump through the air, and lightning flash from thunder clouds.

Losing electrons

You saw in the last unit that the centre of an atom, called the nucleus, is held together by a very strong force. But the force holding the electrons onto an atom is much weaker. Electrons can be taken away just by rubbing.

Look at Figure 1. When you rub a balloon with a woollen duster, the duster sticks to the balloon. There is an electrostatic force between the duster and the balloon. Moving electrons can explain this force.

Rubbing removes electrons from atoms in the duster and gives them to the balloon. Electrons carry a negative (−) charge. The balloon becomes negatively charged, the duster becomes positively charged. The + and − charges *attract* each other. These + and − charges are called *static electricity*.

If two balloons are rubbed by a duster they both gain electrons. Figure 2 shows what happens when they meet. The two negative charges push each other apart – they *repel* each other.

Here are the important points about static electricity:
- It is always made when two things rub together.
- One thing gains electrons and becomes negatively charged (−).
- The other loses electrons and becomes positively charged (+).
- Different charges (+ and −) attract each other.
- Same charges (− and −, or + and +) repel each other.

Even the ancient Greeks knew about static electricity. But they could not explain it as simply as this.

van de Graaff generators

You can make electricity by rubbing plastic combs, pens, balloons, sheets of glass, and nylon jumpers. Figure 3 shows an easier way – a machine called a van de Graaff generator, invented in 1929. It uses a moving rubber belt to carry negative electric charge up to a shiny 'dome'. The charge builds up on the outside of the dome. When a metal ball comes close, a spark jumps through the air. The electric charge on the generator is carried away through the air. The generator *discharges*.

You probably know that batteries can be $1\frac{1}{2}$ volts, 6 volts, 9 volts, or more. 'Mains' electricity used at home is about 240 volts. Some van de Graaff machines can generate up to 10 million volts.

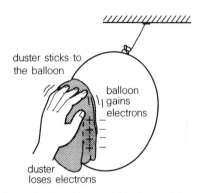

Figure 1 The duster and balloon stick together because there is an electrostatic force between them.

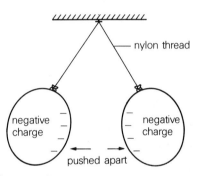

Figure 2 When two objects have the same charge, they repel each other.

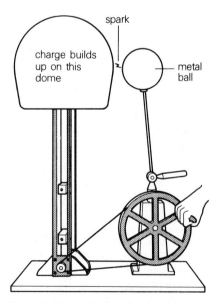

Figure 3 A van de Graaff generator is a machine which makes electricity.

Lightning   A cloud in a thunderstorm carries an enormous amount of electric charge. When it comes close to a tree, a tall building, a chimney, or even a person the electric charge travels through the air as a giant spark. The cloud needn't be very close – some sparks of lightning are 6 kilometres long. The noise of the thunder would take 18 seconds to reach you, if you lived that long.

It takes a van de Graaff generator or a thunder cloud to make electricity or electric charge travel through air. But some materials carry electricity much more easily . . .

Moving electrons

The word 'static' means 'standing still'. The static electricity on balloons, plastic combs, and nylon clothes is electricity that is not moving. Materials like plastic and nylon do not *carry* electricity – they are called *insulators*.

Conductors and insulators   Figure 4 shows a simple way of finding out which materials carry electricity, and which ones don't. If a piece of copper wire is placed between the two clips the bulb lights up. The copper must be carrying electricity through it. With a plastic comb instead of the copper wire, the bulb will not light. Plastic will not carry electricity.

Anything which carries electricity is called a *conductor*. Some liquids are good conductors of electricity, as Figure 5 shows. Salty water and mercury (the only 'liquid metal') are both carriers or conductors.

The table below lists substances as those that conduct electricity very well, those that conduct electricity a little, or those that do not conduct at all – insulators. You can see that some conductors are liquids, some are solids. Air will carry electricity when it is damp, but not when it is dry. This is why experiments with static electricity work best on a dry day – the dry air will not carry the 'static' away.

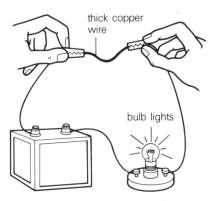

Lightning is a huge spark of electricity travelling through the air.

good conductors	conductors, but not very good	insulators
all metals,	human bodies	dry air
including	frog's legs	plastic
copper	dirty water	balloons
silver	salty water	nylon socks
gold	damp air	wood
mercury	sweaty hands	rubber
	cows on electric fences	distilled water
carbon	silicon	

Figure 4   The bulb only lights up if a conductor is placed between the metal clips.

How do conductors carry electricity?   Static electricity can be made by rubbing electrons loose from their atoms. The electricity stands still (is static) because these loose electrons cannot travel through an insulator such as plastic. But a conductor will carry these loose electrons. A copper wire is a good conductor because it lets electrons travel through it.

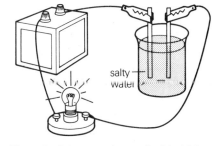

Figure 5   Salty water is one liquid which will conduct electricity.

Electric circuits

Electrons travelling through a conductor make an *electric current*. In the 18th century scientists thought that electricity was a kind of fluid, like water, that flowed through wires. Now we believe that electric current is a *flow of electrons*.

Pushing electrons Heat travels or flows through a metal bar from the hot end to the cold end. In a similar way electricity can travel through metal – but it must be pushed. Electrons can be pushed through a bar by a battery, as Figure 6 shows. Every battery has two connections: a positive (+) and a negative (−). The electrons are pushed *away* from the negative side of the battery *towards* the positive side.

Pathways Like cars, electrons always need a pathway to travel along. They always travel in a *circuit*. When they reach a 'dead end' – when the circuit is broken – they stop, as Figure 7 shows. This is how you switch a light off, by stopping electrons from reaching the bulb.

Wide and narrow pathways Cars can travel on rough or smooth roads, wide or narrow roads. In the same way some pathways for electric currents are better than others. Thick wires make a better pathway than thin ones. Just as wide roads carry more traffic, thick wires can carry more electrons. Figure 8 shows that, like cars, most electrons take the easiest route.

Remember that moving electrons are like 'cars of electricity':
- they always need a pusher
- they must have an unbroken pathway
- they prefer the easiest route.

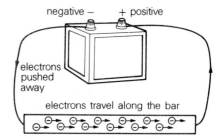

Figure 6 Electrons can be pushed through a metal bar by a battery.

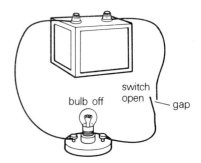

Figure 7 If a circuit is broken, electrons cannot flow.

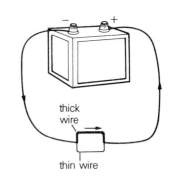

Figure 8 Electrons will travel along a thick wire in preference to a thin wire.

Summary

1 Static electricity is made by rubbing electrons loose from their atoms.

2 There are two kinds of electric charge: + and −.

3 If enough charge builds up on an object it can travel through the air as a spark. Flashes of lightning are huge sparks carrying electrons (− charge) through the air.

4 Materials that carry electrons are called conductors.

5 Materials that do not carry electrons are called insulators.

6 Electrons travelling through a conductor make an electric current. Every current needs both a 'pusher' and a 'pathway'.

Exercises

1 Give three examples of static electricity.

2 Copy out and fill in the blanks:
Two positive charges _____ each other. A positive and a negative charge _____ . The word 'static' means _____ _____ . Large amounts of static electricity can be made with a _____ _____ _____ generator. _____ is made by giant electric sparks in a storm.

3 What is the difference between a conductor and an insulator? Give two examples of each.

4 a) What is an electric current?
   b) What is an electric circuit?

5 Explain why:
   a) a balloon becomes negatively charged when you rub it with a duster
   b) some materials carry electricity but others don't
   c) you see the lightning in a storm before you hear the thunder
   d) a light switch must be on before the bulb will light.

5.3 Electric circuits

Electricity always needs a 'pusher' and a 'pathway'. The pusher can be strong or weak; the pathway can be long or short, easy or difficult. This unit tells you more about the pathways that electricity can take and how you can draw them quickly.

'Drawing' electric circuits

Figure 1 shows a very simple electric circuit connected together on a special wooden board called a *circuit board*. The circuit uses a single cell to push electricity through a small light bulb. When the switch is pressed down the pathway or circuit is complete, electrons flow, and the bulb lights. Electricity always needs a complete pathway to travel along.

Circuit diagrams There is a much quicker way of showing electric circuits than by drawing each part. You can use symbols to stand for each bit of the circuit:

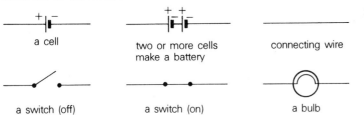

a cell   two or more cells make a battery   connecting wire

a switch (off)   a switch (on)   a bulb

When these symbols are joined together they make a *circuit diagram*. They are rather like maps that use symbols to stand for different objects (e.g. ŏ for a church). The circuit diagram in Figure 2 is a map of the circuit on the board in Figure 1. These circuit diagrams or maps will be used all through this unit.

Although electrons actually travel from − to +, circuit diagrams usually show the current going the other way. This direction was agreed by convention before electrons were discovered. It is called 'conventional current'.

Measuring electric current

Electric current is measured in *amperes*, named after a French physicist called André Ampère (1775–1836). The current in a small light bulb is only about one-quarter of an ampere ($\frac{1}{4}$ A for short), in a mains bulb about $\frac{1}{2}$ A, and in an electric fire about 4 amperes (or 4 amps).

You can get a rough idea of the size of an electric current by seeing how brightly it lights a bulb. But current can be measured more accurately using a special meter called an *ammeter* (short for 'ampere meter'). Figure 3 shows an ammeter connected into an electric circuit. Like a cell, it has two connections: a positive (+) which is red, and a negative (−) which is black. The ammeter must be connected the right way round, or the needle will move backwards instead of forwards.

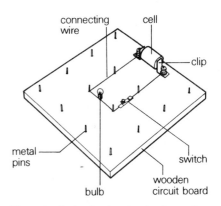

Figure 1 A simple electric circuit can be built on a circuit board.

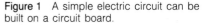

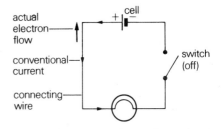

Figure 2 This circuit diagram is a map of the circuit in Figure 1.

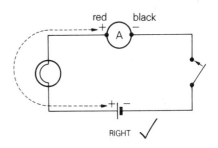

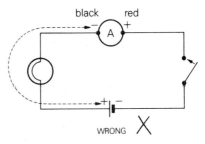

Figure 3 The red connection on an ammeter must be joined to the positive connection of the cell.

Increasing the current

How can you make the current in a simple circuit bigger? The easiest way is to use two cells instead of one, as Figure 4 shows. This doubles the driving force pushing the electric current around the circuit. The current becomes twice as big as before, and the bulb gets much brighter. Figure 5 shows a circuit with three cells and only one bulb. There are three "bulb's worth" of current going through one lamp. This is too much for it – the filament wire inside the bulb melts. The bulb blows.

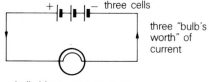

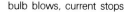

bulb blows, current stops

Figure 5 Adding more cells increases the current – sometimes too much!

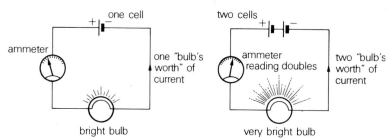

Figure 4 Increasing the number of cells increases the current.

Voltage The driving force or electrical push in a circuit is called the *voltage*. It is measured in **volts** with a special meter called a **voltmeter**. Figure 6 shows how a voltmeter can be used as a kind of 'cell counter'. Suppose each cell has an electrical push of $1\frac{1}{2}$ volts. If the voltmeter is connected across one cell it reads $1\frac{1}{2}$ volts. Across two cells it reads 3 volts; across three it reads $4\frac{1}{2}$ volts. Voltmeters measure the push driving the current around a circuit.

In 1826 a German called Georg Ohm first said that:
- The larger the voltage, the bigger the current.
But he knew that there are other ways of changing the current in an electric circuit . . .

Resisting the current

Adding more bulbs The current driven by three cells was too much for the single bulb in Figure 5. It blew. The circuit in Figure 7 has two more bulbs added. Each bulb now shines with its normal brightness. The electric current flowing through the circuit is just the right strength to light each bulb. If more bulbs were added the current in the circuit would get weaker and weaker. Six bulbs would make the current so small they would not light.

Each bulb added makes more *resistance* in the circuit. More and more resistance makes the current weaker and weaker.

Georg Ohm first realised that there are two ways of increasing the electric current in a circuit:
- by *lowering the resistance* or
- by *increasing the voltage*.

Thick and thin wires Electricity travels more easily through thick wires than through thin wires. The thin wire inside a bulb *resists* the movement of electrons through it. It has a *large resistance*. Very thick wires or strips of metal allow electrons to travel through them easily. They have a *small resistance*. Look at Figure 8A. The

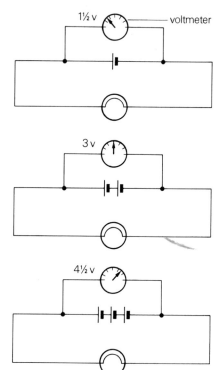

Figure 6 A voltmeter measures the electrical push driving the current around a circuit.

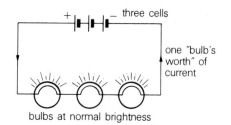

bulbs at normal brightness

Figure 7 Adding more bulbs makes the current weaker.

electrons in this circuit can choose between the thin wire in the bulb and the thick strip of copper. They always take the easiest pathway. They go through the copper strip and the bulb goes out. This is called a *short circuit*.

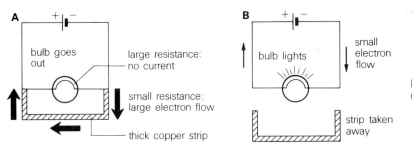

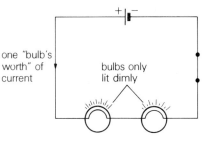

Figure 9 The two bulbs in this circuit are connected in series.

Figure 8 Electrons always take the easiest pathway.

A very large current goes around this easy route and would soon run the cell down. If the copper strip is taken away the bulb lights again. The current is much smaller but it all goes through the bulb.

Different pathways

The current and resistance in a circuit can also be changed by rearranging the pathways in the circuit.

Series and parallel Figure 9 shows a circuit with two bulbs connected in *series*. The cell gives enough driving force to light one bulb brightly. When *two* bulbs are connected in series, each one is only dimly lit. One cell cannot provide enough driving force to light the two bulbs as brightly in series because the total resistance is greater.

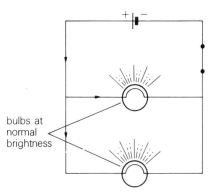

Figure 10 The two bulbs in this circuit are connected in parallel.

Figure 10 shows a circuit with two bulbs connected in **parallel**. Each bulb has the full driving force of the cell across it. The cell pushes enough current through each bulb to make it glow brightly. Each bulb takes one bulb's worth of electric current from the cell. It is as if each one were connected *separately* to the cell. But the cell will be used up twice as quickly.

Where the electrons go Figure 11 shows where the moving electrons go in the different pathways. In the parallel circuit the electric current splits up, some going through one bulb and some through the other. You can prove this by putting three ammeters in the circuit in the positions shown in Figure 12.

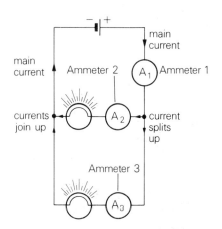

Figure 12 The ammeters show that the current is not the same in each part of the circuit. The bulbs share the current between them. The readings on Ammeter 2 and Ammeter 3 should add up to be the same as on Ammeter 1.

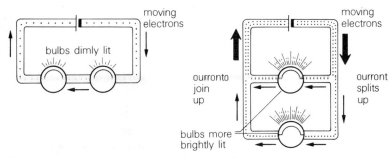

Figure 11 In the parallel circuit the current splits up.

In the series circuit, the electrons do not split up. The current is the same throughout the circuit. Three ammeters placed at three different places in the circuit all give the *same* reading. None of the electric current is used up or lost as it goes through the bulbs. But there is *more* resistance when there are two bulbs in series than when there are two in parallel. So the current in the series circuit is *less* and each bulb is dimmer than in the parallel circuit.

Using series and parallel   Both kinds of pathway can be used to make electric circuits. Light bulbs can be connected either in series or parallel. In series if one bulb blows, or gets broken, all the other bulbs go out. The pathway is broken and the current cannot flow. But in parallel if a bulb blows, or is broken, the other bulbs stay on. They still have a complete pathway. Christmas tree lights can be connected in parallel as Figure 13 shows. Then if one goes out the other lights stay on.

good conductor, large current
good road, large volume of traffic

bad conductor, smaller current
bad road, less traffic

insulator, current stops
barrier, traffic stops

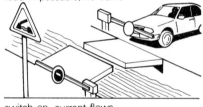

switch off, no current
road impassable, no traffic

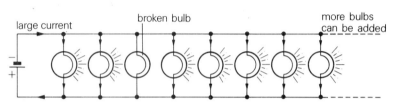

Figure 13   In parallel, a broken bulb does not affect the other bulbs.

switch on, current flows
barrier removed, traffic flows

Figure 14   Moving electrons can be compared with cars.

Summary

1   Electric circuits can be drawn by using symbols to stand for the different parts.

2   Electric current is measured in amperes, with an ammeter.

3   Voltage is measured in volts, with a voltmeter.

4   There are two ways of increasing the current in a circuit – by lowering the resistance or by increasing the voltage.

5   Parts of a circuit can be connected in two ways, called series and parallel.

6   Figure 14 shows how you can compare moving electrons in circuits with moving cars on different roadways. But unlike cars, electrons do not have their own driving force – they need to be pushed by a cell, battery, or magnet.

Exercises

1   Draw a diagram of an electric circuit containing: a cell, a switch, and two bulbs.

2   Draw an electric circuit with a battery and:
a) three bulbs in series
b) three bulbs in parallel.
Which circuit uses up the battery more quickly?

3   Which arrangement is best for Christmas tree lights: series or parallel? Explain why.

4   Describe what is meant by a 'short circuit'.

5   Draw a simple diagram to explain what is meant by 'conventional current'.

6   Copy out and fill in the blanks:
Electric current is measured in _____ , named after a Frenchman called _____ _____ . It is measured by an _____ . The voltage in a circuit is measured with a _____ . In _____ a German called Georg _____ first said that the larger the _____ the bigger the current in a circuit.

7   Explain, in your own words, what an electric current *is*.

8   Make a table showing all the circuit symbols you know of, and what each one stands for.

5.4 Electricity at home

Power pylons carry electricity at high voltage.

The electricity you use at home is called 'mains' electricity – it has a driving force of 240 volts. Where does it come from? How does it reach the plug on your wall? Why is it so useful? How can it be used safely? This unit tells you about the electricity you use at home.

From power station to plug

The electrical energy supplied to homes and factories is generated in huge power stations. Very thick cables are used to carry electrical energy from these power stations to houses, offices, and factories. A special instrument called a **transformer** is used to change the voltage of this electricity. The thick overhead cables in Figure 1 carry electricity at many thousand volts. Before it enters a house its voltage must be brought down to 240 volts by a transformer.

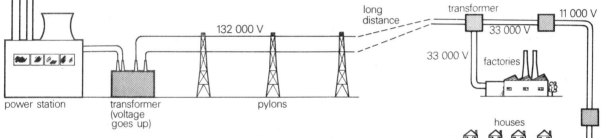

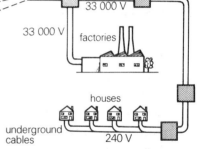

Thick cables bring the mains electricity into the house. Once the electricity is inside the house it goes through a special **meter** that measures how much energy you use. Then it goes through the main fuse box, as Figure 2 shows. From this box several cables carry electricity to different parts of the house for lighting, heating, cooking, etc. Each cable has its own fuse inside the fuse box.

Figure 1 Electrical energy is carried from a power station to a house or factory in thick cables.

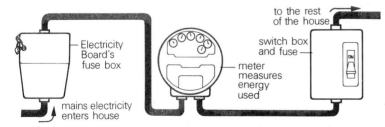

Figuro 2 Maino olootrioity muot paoo through a motor and a fuoo box.

The ring main The cables to the power sockets and plugs are laid in a ring around the house. This ring starts *from* the main fuse box, runs around the rooms of the house to each of the sockets, and back again *to* the fuse box, as Figure 3 shows. Each cable from

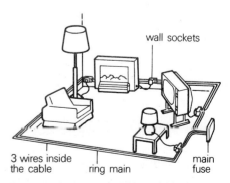

Figure 3 Cables to the power sockets are laid in a ring around the house, starting and ending at the fuse box.

127

the fuse box has three separate wires inside it, each covered in insulation. The three wires are called the *live, neutral,* and *earth* wires, as shown in Figure 4. Each of these wires is connected into every socket on the ring.

The earth wire   The third wire inside the cable running to each socket is joined to a metal plate or rod which is buried in the Earth. This is called the earth wire. Normally it does not carry any electric current – it is there for *safety* as the next section explains.

Plugging in

Figure 5 shows a plug on the end of a cable which has been *correctly* wired to go into a power socket. Each wire inside the cable is covered with specially coloured plastic insulation. The insulation on:

- the live wire is brown
- the neutral wire is blue
- the earth wire is striped green and yellow.

These wires must always be connected to the right parts of the plug.

Fuses   A fuse is a short piece of thin wire which becomes very hot and melts if too much current goes through it. Figure 6 shows a simple fuse. As the fuse melts or 'blows' it breaks the electric circuit and the current stops.

Fuses and switches must always be placed in the *live* wire. If they were placed in the neutral wire the instrument being used would still be live, even with the switch off or the fuse blown.

One fuse is placed in the main fuse box for each 'ring main' that comes from it. Every plug should also have its own fuse (next to the live wire) which blows if the current suddenly becomes too large and dangerous.

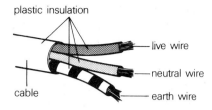
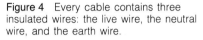

Figure 4   Every cable contains three insulated wires: the live wire, the neutral wire, and the earth wire.

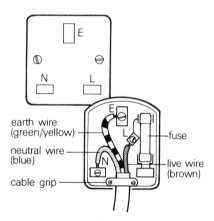

Figure 5   This plug has been wired correctly.

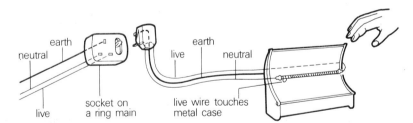

Figure 7   An electrical device should be connected to an earth wire.

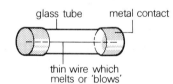

Figure 6   A fuse is a short piece of thin wire. If too much current goes through it, it melts and breaks the electric circuit.

Earthing   Figure 7 shows a plug for an electric fire going into a power socket. Suppose this fire became damaged and the live wire touched the metal casing. The metal case would now be 'live'. If someone touched it, and there was no earth wire, he or she would get a nasty shock. So anything with a metal case should be connected to an earth wire. If the case becomes live the electricity takes the easiest path – through the earth wire. A large current travels along this wire and safely down into the Earth. This large current blows the fuse.

This fuse has a fuse value of 13A. If a current of more than 13 amps flows through it, the fuse blows.

Fuse values Different electrical devices need different sized fuses in their plugs. The fuse must always be able to take *slightly more* current than the device actually uses, otherwise it would melt too easily. The size of a fuse is called the **fuse value**. Common fuse values are 3A, 5A, and 13A. If a kettle normally takes a current of 10 amps (10A), its plug needs a 13A fuse. A TV that uses a current of 1A needs a 3A fuse. Other examples include:

fuse value	device
3A	TV, table lamp, record player, electric blanket
5A	vacuum cleaner, electric iron, small electric fire, hair drier
13A	kettle, large fire, immersion heater

The fuses in the main fuse box are much larger and can carry a bigger current. Fuses for power ring mains are 30A fuses; for lighting 5A fuses are used. These will blow if the current going around the ring main in the house is too large.

Using electrical energy

Whenever you 'plug in' and use electricity at home you are changing electrical energy to another form of energy.

Heaters and lighters Thin wires have a high resistance and get very hot when carrying an electric current, as Figure 8 shows. The current from three cells can make this thin wire so hot it will cut through a piece of polystyrene. Electricity from the mains can be changed into heat energy in long thin wires. Figure 9 shows a few examples. The coil of thin wire in each appliance is called a *heating element*. It changes electrical energy into heat energy.

These coils of thin wire become red-hot at about 900 °C. The coil of thin wire in a light bulb becomes even hotter (about 2500 °C). It becomes white-hot, and gives out light energy as well as heat. Why doesn't the wire melt, or even burn? As Figure 10 shows the filament wire inside a light bulb is surrounded by a special gas called argon. The filament will not burn in argon. The filament is made from a special metal called tungsten which can become white-hot without melting.

Moving energy Many of the things people use at home change electricity into movement: electric drills, spin driers, washing machines, mixers, fans, and so on. They all use electric motors to change electrical energy into kinetic energy.

Sound energy Electric bells and loudspeakers change electrical energy into sound energy.

Power

Different electrical devices need different fuses in their plugs. Some use small currents, less than 1A, and need a 3A fuse. Some use much larger currents, up to 10A or more, and need a 13A fuse. The more current a device needs, the more energy it uses every second. The energy used every second is called *power*. This is

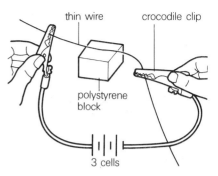
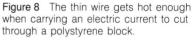

Figure 8 The thin wire gets hot enough when carrying an electric current to cut through a polystyrene block.

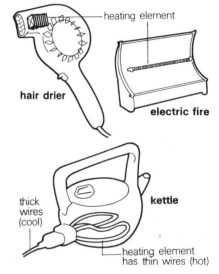

Figure 9 The heating element in each appliance changes electrical energy to heat energy.

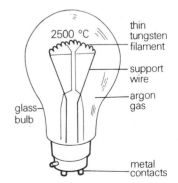

Figure 10 The tungsten filament in a light bulb glows white-hot when carrying an electric current.

measured in *watts*. The power of a light bulb may be 60 watts (60W, for short). This means that it changes 60 joules of electrical energy every second into another form. The power of an electric kettle may be 2400 watts or 2.4 kW. The kettle converts 2400 joules of electrical energy into heat energy every second.

These are only rough values. Most electrical devices have an 'information plate' on them telling you exactly how many watts they use. An example is shown in Figure 11.

Look at the examples listed in the table on the next page. Notice that the larger the current (in amps), the larger the power (in watts). There is a simple formula connecting power and current:

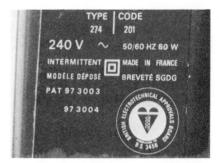

Figure 11 An 'information plate' tells you how many watts an electric appliance uses.

> power = voltage × current
> (in watts)   (in volts)   (in amps)

The power of each device shown in the table can be worked out by multiplying the current by 240 volts, the mains voltage.

Paying for electrical energy

Devices that use a lot of current have a larger power and use electrical energy quickly. They cost more money to run. Electric kettles, cookers, fires, and immersion heaters use more energy every second than light bulbs, electric clocks, and televisions.

The electricity bill   Look closely at your electricity meter at home. It is marked in *kilowatt-hours* (kWh, for short). With this meter the Electricity Board measures the electrical energy you use. For example: *One* kilowatt for *one* hour uses *one* kWh of energy

An electricity meter measures how much electricity is used.

   Two kilowatts for *one* hour uses *two* kWh of energy

   Two kilowatts for *two* hours uses *four* kWh of energy.

You can use this simple formula:

> energy used = power × time
> (in kWh)   (in kW)   (in hours)

So a 3 kW electric fire, left on for 4 hours, uses 3 × 4 or 12 kilowatt-hours of energy.

Paying for it   One kilowatt-hour is called one *unit* by the Electricity Board. Suppose each unit you use costs 6p. If you leave a 2 kW heater on for 2 hours, it uses 2 × 2 or 4 kWh of electrical energy. This will be '4 units' on your bill, and will cost 4 × 6p or 24 pence.

Saving it   The larger the power of a device, and the longer you leave it on, the more energy it uses, and the more it costs. Everyone needs to make an effort to save electrical energy – not just to save money, but also to save the fuels like coal and oil that are burnt to generate electricity at power stations.

The following table shows you the current, the power, and how long it takes for different devices to use 1 unit of electrical energy:

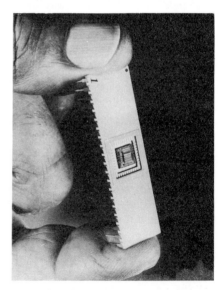

The availability of electrical energy revolutionised the number of machines used in everyday life. A second revolution occurred with the development of computers. Many machines used in the home today contain a microprocessor like the one in this photo.

device	current (in amps) (roughly)	power (in watts) (roughly)	how long to use 1 unit? (roughly)
electric clock	0.01 A	3 W	333 hours (about 2 weeks)
light bulb	0.25 A	40 W, 60 W, 100 W	10 hours
stereo system	0.5 A	110 W	9 hours
fridge, electric blanket	0.5 A	120 W	8 hours
television (B. & W.)	0.5 A	120 W	8 hours
(colour)	1.5 A	320 W	3 hours
drill	1.5 A	360 W	3 hours
vacuum cleaner	2 A	500 W	2 hours
hair drier	2.5 A	600 W	1½ hours
electric iron	4 A	960 W	1 hour
electric fire or kettle	8 A	2000 W (2 kW)	half an hour
tumble-drier	9 A	2200 W (2.2 kW)	25 minutes
cooker	16 A	4000 W (4 kW)	15 minutes

Summary

1 Electrical energy is carried to your home by very thick cables from power stations. It travels through a meter and a fuse box before it is carried around the house by smaller cables in a ring main.

2 Plugs have three connections: live (brown), neutral (blue), and earth (green and yellow). The earth wire is for safety and carries a dangerous current away to the Earth. This current blows the fuse in the plug.

3 All electrical devices change electrical energy to another useful form of energy: light, heat, sound, or kinetic energy.

4 Devices that use a large current use electrical energy very quickly. They have a large power, measured in watts.

5 The electrical energy you use is measured by a meter in kilowatt-hours. One kWh is called one unit.

This nuclear power station in Scotland supplies over one thousand million units of electricity to the National Grid each year.

Exercises

1 Describe how electrical energy gets from power stations to the plugs in your house. What is the size of the 'mains voltage'?

2 What are the three wires inside a power cable? Draw a diagram to show how each of these wires should be connected to a plug. Show where the fuse goes.

3 What value of fuse is best for:
a) a record player?
b) an electric kettle?
c) a TV?
d) a vacuum cleaner?

4 Describe three different uses for electrical energy.

5 What is the power (roughly) of:
a) an electric clock?
b) a colour TV?
c) an electric fire?

How much energy does each one use in:
a) one second?
b) one minute?
c) one hour?

6 Explain why:
a) fuses and switches are placed in the live wire
b) instruments with a metal case should be earthed
c) every plug should have a fuse.

Nuclear Power: Yes Please or No Thanks?

. . . that is the question. Should nuclear energy be used to supply our electricity in the future? Why are so many people against it? After all, the Sun's energy comes from a nuclear reaction.

All those in favour . . .

A nuclear power station uses the heat from splitting atoms, just as a coal-fired power station uses the heat from burning coal. Britain built the world's first nuclear power station in 1956. There

Calder Hall was the world's first nuclear power station.

are nearly 20 stations around the coasts of Britain. One kilogram of pure uranium can release as much heat energy as two hundred tonnes of coal!

The people in favour of nuclear power believe that nuclear energy is safe, clean, and cheap. It seems almost too good to be true . . .

All those against . . .

Is nuclear power safe, clean, or cheap?

Safe? If a nuclear reactor went wrong the inside could become very, very hot. The whole reactor could *melt* – this would be called a *melt-down*. The radioactive material inside would leak into the Earth and its atmosphere. Some people believe that a nuclear meltdown could happen and thousands of people would die from radiation 'sickness'. Others say that the chances of a nuclear meltdown are a million-to-one against.

Accidents have happened in nuclear power stations. In 1957 there was a fire in the Windscale nuclear reactor in Cumbria – no one was killed. In 1979 there was almost a meltdown at the Three Mile Island reactor in the USA – the reactor has never been used since.

Then in 1986 a nuclear reactor at Chernobyl in Russia became overheated and caught fire. At least thirty people died in trying to put the fire out. But, more importantly, large amounts of very radioactive material were released into the air. A cloud of radioactive dust was slowly blown across Europe – the dust affected Sweden, Poland, Germany, and Norway. Some people believe that the radiation from Chernobyl even reached the British Isles, over a thousand miles away. Many people became much more concerned about nuclear power after the Chernobyl disaster. Some believe that it could happen again – but many experts say that an accident like Chernobyl could never occur in Britain.

The damaged reactor buildings at Chernobyl in the Soviet Union.

But there may be a far greater danger with the spread of nuclear power: the *bomb*. Reactors use uranium. But while they are working, some material inside the reactor is changed into plutonium. And plutonium can be used to make nuclear weapons. So any country with its own nuclear reactor might be able to make its own nuclear bomb. Many people believe that as more and more countries get their own nuclear reactor the danger of nuclear war will increase.

Clean? Is nuclear power clean? If there are no accidents it will not pollute the air, as burning coal, oil, and petrol does. But the *waste* from nuclear reactors is deadly. Reactors change uranium into other chemicals. These chemicals, the reactor waste, are poisonous and radioactive. The dangerous waste stays radioactive for thousands of years.

This train is loaded with flasks containing radioactive spent nuclear fuel. The flasks are being taken from Barrow to Sellafield where they will be reprocessed.

It's a problem that won't go away. There are only two ways of getting rid of nuclear waste: burying it under the sea-bed or dumping it deep inside the Earth. But will it escape in the next 1000 years? Nobody is really sure.

Cheap? Once a nuclear power station has been built it produces electricity quite cheaply, though not always as cheaply as modern coal-fired power stations. And it costs over £1000 million to build one. A reactor also needs to be closely guarded. Terrorists have to be stopped from stealing plutonium!

The fast reactor centre at Dounreay, Scotland. The reactor is in the foreground.

Another problem is fuel. Only a few countries have their own uranium. One day it will run out, just as coal and oil will. But recently scientists have designed a new reactor, called 'the fast-breeder reactor'. This actually makes its own fuel, plutonium. It may even be able to use up some of the radioactive waste. But any country with a fast-breeder reactor will have plenty of dangerous plutonium and the potential to make plenty of bombs.

This photo shows breeder fuel elements which will be used in the fast reactor at Dounreay. They are manufactured at Springfields, near Preston.

No thanks or yes please

Nuclear power is *not* completely safe, clean, or cheap. Could it solve the world's energy problem? Not on its own. Electricity is good for doing certain jobs: lighting, running electric motors, running televisions, and so on. But it is a very wasteful way of heating buildings. Home heating from nuclear power will never be cheap. And almost half the energy used in Britain goes on heating houses, factories, and offices, and making hot water.

Most of the world's energy ends up as useless heat energy that escapes into the air. The answer is not to make more and more energy, but to *save* the energy we've got. As one expert said: 'If the bath water is escaping, is it more sensible to turn on the taps faster, or to get a better bath-plug?'

One day nuclear power may be completely safe, clean, and cheap. But it is not yet. The money needed to build *one* nuclear power station would pay for insulating the house of *every* old-age pensioner in Britain.

Topic 5 Exercises

More questions on atoms and electrons

1 What are the different parts of an atom called? Find out who discovered each of these parts.

2 Here is a list of different materials: stone, brick, copper, plastic, wood, sea-water, air, pure water, dilute sulphuric acid, nylon, rubber.
Put them in a table to show *conductors* and *insulators* of electricity.

3 This electric circuit has a battery and three identical bulbs. Which bulb is brightest? What would happen if bulb B were removed? What is this type of circuit connection called?

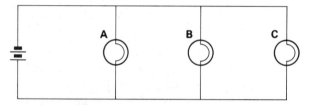

4 Why would it be dangerous to put a switch, or a fuse, in the neutral wire instead of the live wire in an electric circuit?

5 Explain why the filament of a light bulb gets very hot but the flex leading into it does not.

6 Write down five uses for electric motors.

7 Some electrical devices are not 'earthed' – there are only two wires leading into them (e.g. an electric razor). Why is it safe to hold them? Give more examples.

8 How could you estimate the length of the filament in an electric light bulb without breaking the bulb?

Things to do

1 You need: a plastic comb, a tap, something woollen, a ping-pong ball.
   a) **Attracting water**
   Turn the tap on slowly so that a thin stream of water comes out. Then rub the comb on something woollen, a sweater sleeve for example. Bring the comb close to the water. The stream bends towards the comb. By rubbing the comb you have produced static electricity on it. (Instead of a comb you can use a plastic ruler or a pen. Remember that experiments with static electricity work best on a clear, dry day.)
   b) **Pulling a ping-pong ball without touching it**
   Rub the comb again on something woollen. Bring the charged comb close to a ping-pong ball on a table. The ball is attracted by the comb. If you move the comb away the ball rolls after it.

2 Conducting pencils
Most experiments with electricity need equipment. Mains electricity is dangerous and should never be experimented with. But some experiments can be done with wires, bulbs, and batteries.

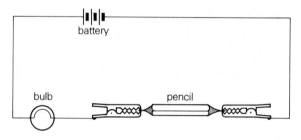

Make an electric circuit with a battery, wires, and a small bulb. Then sharpen a pencil at *both* ends. Place the pencil in the circuit by connecting wires to both ends. What happens? Try connecting different pencils, with different 'hardnesses' (2B, 3B, HB). Is there a difference in the brightness of the bulb? What happens if you use pencils of different lengths?

3 Tricks with tights
You need: an old pair of tights, two short sticks, and a balloon.
Cut one leg off the tights to make a stocking. Wedge the stick into the top of it, then let it hang.

Now rub the blown-up balloon up and down the stocking. The stocking 'fills out' as if it contained a leg. Do the same with the other leg of the tights.

Then bring the two legs together. What happens?

Now bring the balloon near the two legs. What happens now?

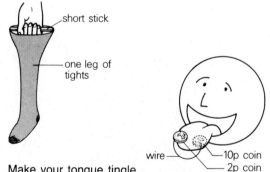

4 Make your tongue tingle
You need: a clean 10p coin, a clean 2p coin, a wire.
Place the two coins on each side of your tongue. Join them with a wire. What do you feel? The two coins, and your tongue, form a simple electric cell.

The rings of Saturn consist of particles of dust and rock held in orbit by Saturn's gravitational field.

6.1 Magnets and magnetic fields

Why do all objects fall to Earth when they are dropped? How can a magnet pick up pieces of iron and steel without touching them? For centuries people have been trying to explain these invisible forces. In 1845 a man called Faraday came up with one useful idea: the idea of 'fields of force'.

What is a field?

Most of the forces in Topic 3 involved things that were *touching* each other. But some forces involve things that are *not touching*, as Figure 1 shows. The magnet is not touching the nail, yet something pulls the nail towards it. The force is invisible – it can be felt, but not seen. There must be a region around the magnet where any iron nail will be pulled towards it. This region is called a *field of force*, a *field* for short.

Types of field   There are three types of field. The space or region around a magnet is called a *magnetic field*. Anything made of iron or steel will feel a force when it is inside this field. All around the Earth there is a huge *gravitational field* that extends far out into space. Anything inside this field will be pulled towards the Earth by gravity. The Earth's gravitational field extends beyond the moon, which is nearly 400 000 km away. The third kind of force field is the *electrostatic field*. A balloon or a plastic comb has an electrostatic field around it when it has been rubbed and charged up. The force from this field will pick up paper, or bend a stream of water.

Remember these three things about a field:
- A field is the region or space within which a force acts. The field can be around a charged object, a magnet, or a large mass.
- It cannot be seen but its presence can be felt.
- It gets weaker and weaker as you move away from the source.

The rest of this unit is about magnets and the fields around them.

Forces between magnets

Magnetic poles   If a magnet is allowed to swing freely it always stops with one end pointing roughly towards the Earth's north pole. This end is called the *north-seeking pole*, or *North pole* for short. The other end points towards the Earth's south pole. This end is called the *south-seeking pole*, or *South pole* for short. The force field of a magnet is strongest at these two ends.

Attracting and repelling   Figure 2 shows a magnet on the end of a piece of cotton. When the North pole of another magnet comes near to the North pole of this one it swings *away*. It is *repelled*. If the South pole comes near it, the magnet swings *towards* the other one. It is *attracted*. This is the important rule about magnets:

> The same poles repel each other; different poles attract.

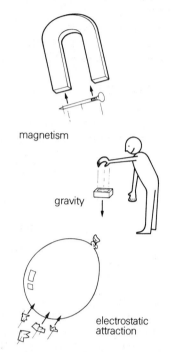

magnetism

gravity

electrostatic attraction

Figure 1   The forces of magnetism, gravity, and electrostatic attraction involve things that are not touching.

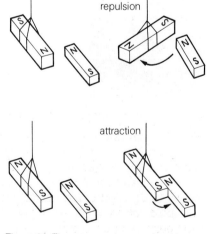

repulsion

attraction

Figure 2   The same poles repel each other. Different poles attract each other.

Picking things up Magnets will pick up or attract certain metals, but not others. Iron, steel, nickel, and cobalt, are attracted by magnets. Copper, brass, and lead are not. Figure 3 shows a magnet picking up an iron nail. The iron nail itself becomes a small magnet, with its South pole joining the North pole of the large magnet. A magnet can be used to pick up a chain of iron paper clips, as Figure 3 also shows. Each paper clip becomes a small magnet. North poles face South poles all along the chain. But when the magnet is taken away the chain collapses – the paper clips are not magnets any longer.

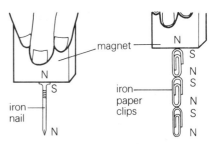

Figure 3 An iron nail and iron paper clips can become small magnets.

Magnetic fields

In 1845 a Londoner called Michael Faraday said that the strength of a magnet is not just *inside* the magnet itself. Its strength is in the space or region *around* the magnet.

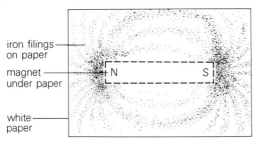

Figure 4 shows some small pieces of iron, called iron filings, on a piece of white paper. With a magnet under the paper the iron filings are arranged into a pattern by the magnet's field of force. Faraday was the first man to call this pattern a field.

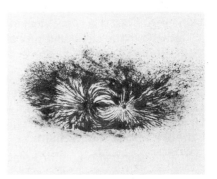

Figure 4 The drawing and photo show how iron filings form a pattern due to the magnet's field of force.

Plotting a field You can draw a magnetic field using a tiny pointed magnet called a *plotting compass*, shown below. When it is near a magnet the compass always points in a particular direction, because of the force of the magnetic field. Figure 5 shows how a plotting compass is used to draw the pattern of this field around a magnet. The North pole of the plotting compass always points *away* from the North pole of the large magnet, *towards* the South pole.

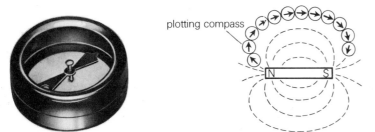

Figure 5 A plotting compass is used to draw a magnetic field.

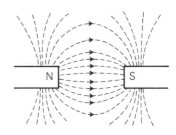

Figure 6 When the two poles are different, the field lines join up.

Fields between two magnets The magnetic field in Figure 5 is drawn with *lines* on it, pointing away from the North pole. What happens to these lines when they come close to another magnet? Figure 6 shows how the lines *join up* when the North pole of one magnet is attracted to the South pole of another one. But when a North pole comes close to another North pole the lines of the two fields push each other apart. They 'spring away' from each other, as Figure 7 shows.

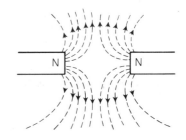

Figure 7 When the two poles are the same, the two fields push each other apart.

The Earth's magnetic field   No-one is quite sure why, but the Earth itself has a large magnetic field all around it. It is the Earth's magnetic field that makes a compass point in a certain direction, like the plotting compass in Figure 5. If you hold a magnet on the end of a loop of cotton its North pole always points towards the Earth's North pole.

Making and destroying magnets

The first people to know about magnetism were probably the Chinese. Over a thousand years ago they found that a special kind of rock, called lodestone, attracted smaller pieces of iron.

Using a magnet to make a magnet   You can stroke a piece of steel with a magnet to make another magnet. Figure 8 shows how. The magnet is rubbed along the steel in *one* direction only, then lifted well away from the steel, and rubbed along again. Eventually the steel becomes a magnet itself.

Destroying magnets   Magnets can be made by stroking gently – they can be destroyed by rough treatment or by hammering them. Magnets also lose their magnetism if they become red-hot.

A theory about magnetism

A large magnet can be cut into pieces, and each piece will be a magnet itself. A magnet can be pictured as made up of lots of very tiny magnets all lined up with their North poles facing the same way. If the magnet is cut into pieces each piece still has tiny magnets inside it, as Figure 9 shows, all facing the same way.

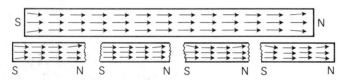

Figure 9   A magnet can be pictured as made up of lots of tiny magnets.

Magnetising   These tiny magnets in an ordinary piece of steel all face in different directions. The steel is *unmagnetised*, as Figure 10 shows. When the steel is stroked with another magnet these tiny magnets are all pulled round so they face in the same direction. The steel is *magnetised*. If the steel is hammered these tiny magnets are shaken up and disturbed. The same thing happens when the steel gets red-hot – the tiny magnets are shaken around by the heat energy until the bar becomes unmagnetised again.

Groups of atoms   Scientists believe that these tiny magnets inside iron or steel are groups of millions of atoms. Each iron atom is itself a very tiny magnet. These 'atomic magnets' in each group line up and all point in the same direction.

Magnetism from electricity

Making an electromagnet   Look at Figure 11. It shows a length of wire wrapped around an iron nail. When an electric current goes through the wire the nail becomes a magnet. It is called an

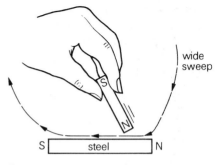

Figure 8   A piece of steel which is stroked with a magnet becomes a magnet itself.

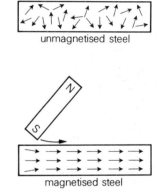

Figure 10   The tiny magnets in unmagnetised steel all face in different directions. In magnetised steel they all face the same way.

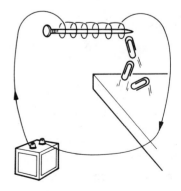

Figure 11   When an electric current goes through the wire the nail becomes an electromagnet.

electromagnet. It will pick up paper clips or iron filings close to it. When the current is switched off the paper clips drop off – the nail is not a magnet any longer.

Inside the nail Figure 12 shows you what is happening inside the nail. With the current off the tiny magnets in the iron all face in different directions. When the current is switched on these tiny magnets line up and point in the same direction. They become 'jumbled up' again when the current is switched off. The iron loses its magnetism. But with a steel nail the tiny magnets carry on pointing in the same direction. Steel keeps its magnetism. Steel is best for making *permanent* magnets. Iron is best for *temporary* ones.

Making a stronger field If a larger current goes through the wire around the nail the electromagnet gets much stronger. Strong electromagnets are used in scrapyards, as Figure 13 shows. They can lift very heavy objects, including cars. They carry a very large current. Smaller electromagnets are used in electric bells, televisions, shavers, cars, telephones, and many other useful devices.

All electromagnets need a supply of electricity. So there must be some *connection* between electricity and magnetism. The next unit is all about this connection.

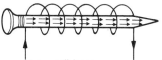

iron nail

current off, nail is not a magnet

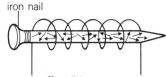

current on, nail becomes a magnet

Figure 12 When the current is on, the tiny magnets in the nail line up and the nail becomes a magnet.

Summary

1 The space all around a magnet is called a magnetic field. A piece of iron or steel in this field will be pulled towards the magnet.

2 Two magnets can attract or repel each other: different poles attract, same poles repel.

3 The shape of a magnetic field can be seen using iron filings or a small compass.

4 A piece of steel becomes a magnet if it is stroked with another magnet. Magnets are destroyed by hammering or red heat.

5 A piece of iron or steel can be pictured as being full of very tiny magnets. These all point in the same direction when the metal is magnetised.

6 Electricity can make a magnetic field. Electromagnets are very useful in everyday life.

Figure 13 Strong electromagnets are used in scrapyards to lift heavy objects.

Exercises

1 Give examples of three different kinds of field. What is meant by 'the field around a magnet'?

2 Why are the two ends of a magnet called the 'North' and 'South' poles?

3 What happens when:
   a) a N-pole meets a N-pole?
   b) a S-pole meets a N-pole?
   c) a S-pole meets a S-pole?

4 Sketch the magnetic field around a bar magnet like this:

   | N   S |

5 Draw the magnetic field between two bar magnets when they face each other like this:

   | S   N | | N   S |

   and then like this:

   | N   S | | N   S |

6 Describe how you can:
   a) make a magnet
   b) destroy a magnet's magnetism
   c) make an electromagnet.

7 Describe the theory of magnetism that scientists use to explain the way magnets behave.

6.2 The electromagnetic connection

Magnets can be made with electricity. Electricity can be made with magnets. There is a strange connection between electricity and magnetism that scientists find very hard to explain. It can be called the electromagnetic connection.

The motor force

When electricity travels through a wire there is a magnetic field all around the wire. Figure 1 shows how this field can be 'seen' using iron filings. The iron filings are made into a pattern by the magnetic field. If a plotting compass is placed near the wire it always points in a certain direction, as in Figure 2. *Every* wire carrying an electric current has a magnetic field around it: this was discovered in 1819 by a Dane called Hans Oersted.

A special force What happens if this wire is placed close to another magnetic field? Look at Figure 3. A wire is free to swing inside a specially shaped magnet called a horseshoe magnet. When you switch the current on this wire has a magnetic field around it. This magnetic field meets or *reacts with* the field from the horseshoe magnet. The wire is pulled *in* or attracted by it. If the battery is turned round the other way, so the current flows the other way, the wire is suddenly pushed *out* when you switch on.

There must be a force pulling the wire in, and pushing the wire out. This is called the *motor force*. It is this force, made when two magnetic fields meet, that is used to drive electric motors.

Motors

There are three ways of making the motor force much stronger:
- by using a stronger magnet
- by passing a larger current through the wire
- by using more than one wire.

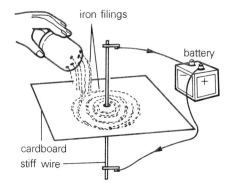

Figure 1 The pattern made by the iron filings shows the presence of a magnetic field when current flows through the wire.

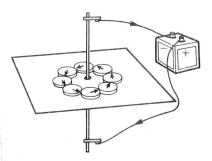

Figure 2 A plotting compass shows the direction of the magnetic field around the wire.

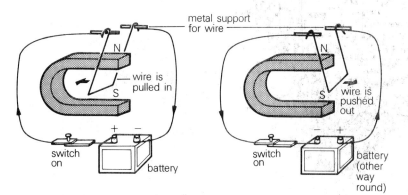

Figure 3 The force between two magnetic fields is called the motor force. This force either pulls the wire in or pushes it out, depending upon the direction of the electric current.

A galvanometer measures electric current.

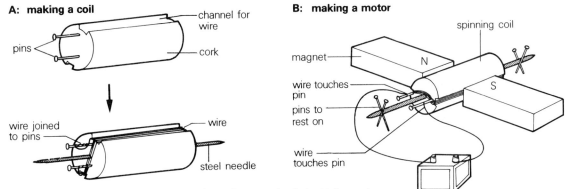

A: making a coil

pins

channel for wire

cork

wire joined to pins

wire

steel needle

B: making a motor

spinning coil

magnet

N

S

wire touches pin

pins to rest on

wire touches pin

Figure 4 To make an electric motor you need a coil, a supply of electricity, and a magnet.

Making a model motor Figure 4A shows how a long wire can be wrapped around and around a piece of cork to make a coil. Each end of the wire is joined to a pin, stuck into the cork. The cork needs a piece of steel through it (like a knitting needle) to hold it up. It can then rest on two supports as Figure 4B shows. Two more things are needed to make a motor: a supply of electricity, and a magnet. The bare ends of two wires from a battery are fixed so that they just brush against the pins on the coil. When the current is switched on it goes through the coil and the magnetic field from the wires meets the field from the magnet. One side of the coil is pushed *up*, the other side is pushed *down*. Give the coil a little flick to start it, and the coil starts to turn. This is a simple electric motor.

Electric motors The electric motors people use need the same parts as this model: a coil of wire, a magnet, and a supply of electricity. Figure 5 shows some of the things that use electric motors. They all have a magnet and a coil inside them. Some use a stronger current than others. Some use a high voltage (from the mains), some use a lower voltage (from batteries). But they *all* change electrical energy into movement.

Figure 5 All these appliances have an electric motor.

Meters

The stronger the current, the stronger the motor force. This rule can be used to measure the size of a current. Figure 6 shows a coil of wire sandwiched between the North and South poles of two magnets. Current goes into the coil, and out again, through two small springs connected to it. When the current is switched on the coil tries to turn, as it does in a motor. But it cannot turn far because it is held back by the springs. The larger the current, the more the coil turns. As it turns it moves a pointer across a scale. This tells you roughly how big the current is.

Moving-coil meters The coil and the magnet in Figure 6 make a simple instrument for measuring electricity. It is called a *moving-coil meter*, or sometimes a *galvanometer* (see photo on page 140). It can be used to detect and measure very small electric currents. The voltmeters and ammeters you saw in Topic 5 are like these meters, but they can be used to measure much larger currents.

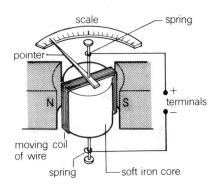

scale

spring

pointer

N

S

+ terminals −

moving coil of wire

spring

soft iron core

Figure 6 A moving-coil meter, or galvanometer, is a simple instrument for measuring electric current.

Electricity from magnetism

You now know that electricity can make magnetism. This section tells you how a magnet can make an electric current.

Figure 7 shows an experiment first carried out by Michael Faraday in 1831. A long coil of wire is connected to a galvanometer. When you push a small magnet into the coil the meter moves – there must be a small electric current going through the wire. When you pull the magnet out of the coil the meter moves the other way. As long as the magnet is moving there is a current going through the wire.

Can a magnet push electrons?   It is almost as if a magnetic field can *push* electrons, just as a battery can push electrons around a circuit. These electrons move when the magnet moves. As they travel through the wire they make an electric current.

Making electric currents   When the magnet is pushed in the electrons move one way – when it is pulled out they move back the other way. This kind of electric current that travels first one way, and then the other, is called an *alternating current*.

The current can be made stronger by:
■ moving the magnet more quickly
■ using a stronger magnet with a stronger field
■ putting more wire into the coil.

This way of making electricity is used in dynamos and generators.

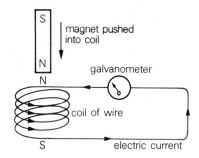

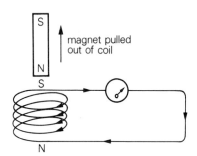

Figure 7   When the magnet moves, current flows through the wire.

Dynamos

Figure 8 shows how a bicycle dynamo uses a magnet moving near a coil of wire. As the wheel goes round it turns a small magnet. The field from this magnet moves through the coil of wire, pushing electrons and making an electric current. This lights the bulb at the front of the bike. The faster you go the quicker the wheel and the spinning magnet turn, the larger the current and the brighter the bulb.

Moving the coil   The bicycle dynamo has a spinning magnet and a coil that stays still. Many dynamos use a spinning coil, inside a large magnet. These dynamos are made exactly like the model motor in Figure 4. A motor *uses* electricity to make the coil turn. But a dynamo *makes* electricity by turning the coil inside the magnet. Dynamos and motors often look the same – they both have the same parts. But a dynamo **makes** electrical energy, while a motor **uses** it. This is summarised in the drawing below.

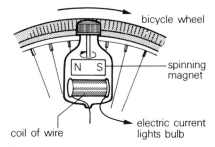

Figure 8   A bicycle dynamo makes electricity from a magnet spinning near a coil of wire.

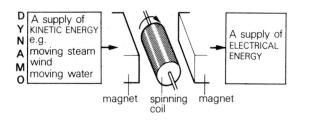

Every dynamo needs energy to make its coil spin. It needs a supply of kinetic energy. A bicycle dynamo uses the kinetic energy of the wheel. Most dynamos and generators in power stations use the energy of hot moving steam pushing a large turbine. The energy of moving water or wind could also be used to turn the coil.

Two types of current

Moving a magnet in and out of a coil makes a current that travels first one way, and then the other. This is an alternating current, *a.c.* for short. With the opposite type of current the electrons flow in one direction only. This is called *direct current, d.c.* for short.

Batteries give a direct current. They always push electrons in the same direction: away from the negative side, towards the positive. Generators can be made to produce either a.c. or d.c. The generators in power stations make alternating current. Mains electricity is a.c.

Both a.c. and d.c. can be used for lighting and heating. But radios and televisions need a d.c. supply – they have a special device inside them which changes the a.c. mains into d.c.

These large turbines are in the Fasnable Power Station at Glen Affric, Inverness. Moving water pushes the turbines, which contain a coil and magnets. Kinetic energy is changed to electrical energy.

Summary

1 Every wire carrying an electric current has a magnetic field around it. When this field meets the field from another magnet there is a force on the wire called the motor force.

2 Electric motors need three parts: a coil of wire, a supply of electricity, and a magnet.

3 Electric meters can be made from a coil of wire inside a magnet. When the wire carries a current the coil turns.

4 Electricity can be made by moving a magnet near a coil of wire.

5 Most dynamos use a coil that is moving near a fixed magnet. The spinning coil generates an electric current.

6 There are two types of electric current. Alternating current travels in both directions along a wire. Direct current travels in one direction only.

A car alternator is an a.c. generator. It contains a moving magnet and fixed coils. The engine turns the alternator and the current generated charges the battery.

Exercises

1 Copy out and fill in the missing words:
When a wire carries electricity there is a _____ _____ around it. This was discovered by Hans _____ . If this wire is between the poles of a _____ a force acts on it. This is called the _____ force. It is used in electric _____ .

2 Sketch a simple electric motor and label: the coil, the poles of the magnet, and the connecting wires. What are electric motors used in?

3 What are the important parts of a moving-coil meter? What is it used for?

4 Explain how a magnet can be used to make electricity. Who discovered this?

5 What is the difference between a dynamo and an electric motor? In what ways are they alike?

6 What are the two types of electric current called? How are they different?

6.3 Gravity and the planets

'What goes up must come down.' How can this well-known 'fact' be explained? Is it true anyway? How do moon rockets escape from the Earth? This unit is all about the pull of gravity. Gravity is another kind of field.

Gravity everywhere: the 'universal force'

There is a story which says that Newton had a brilliant idea after an apple fell on his head. At the time, in about 1665, scientists and astronomers had no idea what force kept the Earth and the planets moving around the Sun. Newton knew that there must be some force pulling the planets around in a circle – otherwise they would travel in a straight line. He also knew that the falling apple in Figure 1 was pulled down by the force of gravity. His brilliant idea was that the same force, *gravity*, extends far out into space. It reaches from the Earth beyond the moon and from the Sun to the furthest planets. The force of gravity extends all through the Universe – it is a *universal force*. With this idea Newton explained the movement of the planets and the moon:

- the Sun's gravity *pulls* the planets in their orbits around the Sun
- the Earth's gravity *pulls* the moon in its orbit around the Earth.

Large masses   Newton realised that you only feel the effects of gravity when very, very large *masses* are involved. 'Every object pulls on every other object with the force of gravity', Newton said. But gravity is about the *weakest* force that we know of. We can only feel it with masses like the Earth's: about 6 million million million million kilograms! The larger the mass, the larger its pull of gravity. The Sun's mass is about 330 000 times bigger than the Earth's. So the Sun's gravity is much stronger than Earth's.

Figure 1   The force of gravity pulls falling objects towards the Earth.

Figure 2   Objects with different masses fall with the same acceleration.

The Earth's gravity

Everyone lives inside the Earth's gravitational field. So whenever you drop something it is pulled towards the Earth. It falls. The first man to study falling objects in a scientific way was an Italian from Pisa, called Galileo. In 1638 he found that two cannonballs of *different* masses both reached the ground at the *same* time if they were dropped from the *same* height. They both fall with the same *acceleration*, as Figure 2 shows.

The atmosphere   The Earth's gravity stops the atmosphere from escaping. Without the atmosphere there would be no life on Earth: it provides oxygen to breathe, protects you from flying meteorites, and holds the Sun's heat in at night.

Escaping from Earth   Most objects thrown or fired upwards into the air come back down to Earth again. But sometimes an object can be fired fast enough to escape completely from the Earth's gravity. To do this it must reach a velocity of more than 11 000 metres per second. This is called the *escape velocity*. A moon rocket must leave the Earth with at least this velocity. The rocket's escape

The Space Shuttle is fired fast enough to orbit the Earth but not to escape completely from Earth's gravity.

144

velocity from the moon will be much less, because the moon's gravity is much weaker.

Orbits and satellites

Sometimes rockets are deliberately fired into the air so that they do not quite escape completely from the Earth's gravity. They are sent into *orbit*, as in Figure 3. They are still inside the Earth's gravitational field, being pulled around in a circle by Earth's gravity. Anything that goes around and around the Earth in an orbit is called a *satellite*. Many satellites are used nowadays to reflect television pictures from one part of the world to another.

Spacecraft Sometimes a spacecraft is sent up so that it orbits the Earth. If the craft has people in it they often have a strange feeling called *weightlessness*. You have the same feeling if you are standing in a lift which suddenly goes down – it feels as if the floor is taken away from under your feet. People in a spacecraft orbiting the Earth have this strange feeling, as if they have no weight. But they are not really weightless – they are still being pulled by the Earth's gravity. A person is only *really* weightless when she is so far away from the Earth and any other big bodies that there is no gravity pulling on her at all.

Moons and planets Our moon is a kind of satellite – it orbits the Earth. Many other planets have moons. Jupiter has at least twelve. In fact planets themselves are a kind of satellite – they are satellites of the Sun . . .

The Sun's family

The Sun and the nine planets which make up our *solar system* are shown in Figure 4. All the planets are inside the Sun's gravitational field, including Pluto which is about 6000 million kilometres from the Sun. They are kept in their orbits by the Sun's gravity. Without it they would fly off into space.

Views of the Universe

Ancient views The Ancient Greeks, in about 600 BC, thought that the Earth was like a flat dish at the centre of the Universe, with the stars, the moon, and the planets going around it. About 700 years later an Egyptian called Ptolemy made up a complicated picture of the Universe with a round Earth at the centre. This picture survived for the next 1400 years, until a Polish monk called Nicolaus Copernicus put the Sun at the centre of the Universe. He said that the Earth was just a planet going around and around the Sun.

Telescopes Shortly after Copernicus died the telescope was invented by a Dutchman in about 1609. Galileo made himself a telescope and used it to view the solar system. Everything he saw supported Copernicus's idea. He also discovered Jupiter's moons and the sun-spots on the Sun. Galileo really believed that Copernicus was right. But the Church in Italy could not accept it – they felt that God must have created the Universe with the Earth at the centre. Galileo was forced to keep his beliefs to himself. (It was not until October 1980 that the Catholic Church and the Pope officially forgave him!)

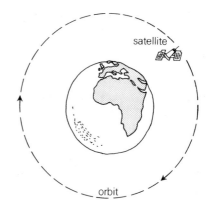

Figure 3 A satellite remains inside the Earth's gravitational field. This means it orbits the Earth instead of flying off into space.

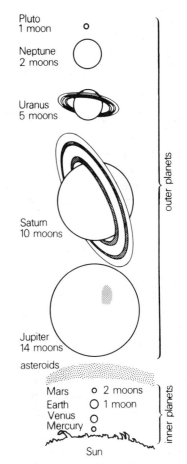

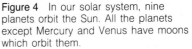

Figure 4 In our solar system, nine planets orbit the Sun. All the planets except Mercury and Venus have moons which orbit them.

145

Newton's theory Newton was the first man to explain *why* the planets were constantly moving around the Sun. His theory still holds today. Since Newton's time the other planets have been discovered: Uranus (in 1798), Neptune (in 1846), and Pluto (in 1930). But they all fit neatly into Newton's great theory.

So far, nobody has explained how the universe actually *began*.

What is gravity?

How Copernicus and Galileo first explained how the universe is arranged – the planets move around the Sun which is at the centre of the solar system.

Why Newton's idea that gravity is universal helped him explain why the planets are continually moving – they are kept in their orbits by the Sun's gravity. Without gravity, they would move in a straight line.

What But *what* is gravity? Can anyone explain this? Everyone knows that it is some kind of 'pulling force': it pulls objects to the Earth, it pulls the moon around the Earth, it pulls the planets around the Sun. But nobody has any idea what it is. In the same way nobody knows what magnetism is. Perhaps these are questions that cannot be fully answered . . .

When there is no gravity a dentist doesn't need a reclining chair!

Summary

1 Isaac Newton realised in 1665 that the force of gravity keeps the planets in their orbits.

2 Gravity pulls all objects towards each other but its effects can only be seen with large masses like the Earth or the Sun.

3 The Earth's gravity makes objects fall and holds the Earth's atmosphere to its surface. Rockets must travel very fast to escape from the Earth's gravity.

4 Satellites orbiting the Earth are pulled around in a circle by the Earth's gravity.

5 A spaceman in orbit feels 'weightless' although he is still inside the Earth's gravitational field.

6 The Sun's gravity keeps the planets of the solar system in their orbits. They are all inside the Sun's gravitational field.

7 Until the 16th century people believed that the Earth was at the centre of the universe. Copernicus, Galileo, and Newton changed that idea.

8 Nobody can explain what gravity really is.

Nicolaus Copernicus

Exercises

1 Copy out and fill in the missing words:
Gravity is another kind of _____ . It extends all through the _____ . The Sun's gravity keeps the _____ in their orbits. Gravity can only be felt with very large _____ .

2 Explain why:
   a) objects fall to the Earth
   b) the atmosphere does not escape
   c) a moon rocket needs to reach a certain velocity.

3 Write down four examples of things that are in orbit.

4 Which planet is:
   a) nearest the Sun
   b) furthest from the Sun
   c) the largest
   d) likely to be the coldest
   e) likely to be the hottest.

5 Explain how scientists' views of the universe have changed from the time of the Ancient Greeks up to the present day.

6.4 Electrons moving in a field

The picture on a television screen is made when electrons strike the screen.

The picture on your television screen is made up of thousands of tiny strips. Behind the screen is a special 'tube'. This unit tells you how the picture is made when tiny invisible particles called electrons travel through the tube, and strike the screen.

Planets and electrons

The planets in the solar system are pulled around in their orbits by the Sun's gravity – they are all inside the Sun's gravitational field. As Figure 1 shows, an atom is a bit like a tiny solar system. An electron in a hydrogen atom goes around and around a proton at the centre. The Sun is rather like the proton, while the Earth in orbit is like the electron.

You cannot see gravity pulling the Earth around the Sun. But it is certainly there. The Earth is inside the Sun's invisible 'field of force'. In the same way an electron is pulled around by the proton's field of force, its *electrostatic field*. This is often called an *electric field* for short. Electric fields can be very useful.

Free electrons Electrons are pulled around the centre of an atom by an electrostatic field. But, unlike planets, they can break away and become *free electrons*.

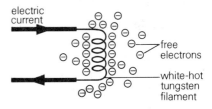

Figure 1 The Earth is inside the Sun's gravitational field. The electron is inside the proton's electrostatic field.

Making free electrons

Hot wires When a piece of wire gets very hot ('white-hot') it starts to release free electrons. A coil of thin wire called a *filament* is shown in Figure 2. It is almost as if electrons are 'boiled off' from the hot filament – they have so much energy at this temperature they break away and become free.

Special tubes Ordinary light bulbs have a white-hot filament inside them. Why don't they provide free electrons? Light bulbs are filled with a special gas called argon which stops most of the electrons from escaping. But if this gas were taken away the electrons would be free to escape. This is just what happens inside a television tube. The tube contains empty space – a *vacuum*. The free electrons can move through the vacuum and make a picture on your screen. The tube is called a *cathode ray tube*.

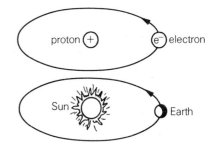

Figure 2 When a filament gets very hot it releases free electrons.

Cathode ray tubes

Figure 3 shows the important parts of a cathode ray tube:
- a hot filament called a *cathode* to provide the free electrons
- a *vacuum* trapped in a glass tube for the electrons to travel through
- a *screen* at one end of the tube coated with a special material that glows when electrons hit it. This is called a *fluorescent screen*
- an *anode* with a very high positive voltage to attract the electrons. The negative electrons from the cathode are drawn towards the positive anode. When they reach the anode they are going so fast that some of them go straight through a special hole in the anode. This makes a narrow beam of fast-moving electrons.

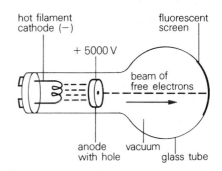

Figure 3 Free electrons in a cathode ray tube make pictures on a TV screen.

This beam of electrons was first called 'cathode rays', when scientists thought they were another kind of ray. Then just before 1900 a man from Manchester called J. J. Thomson said that these rays were really streams of tiny moving electrons. Figure 4 shows the kind of cathode ray tube Thomson used. His discovery led to a whole new branch of physics – *electronics*.

Figure 4 Thomson's cathode ray tube

Bending cathode rays

Using a magnetic field   Normally the stream of electrons in these tubes moves in a straight line, just as a beam of light does. In fact many scientists in the 19th century thought that cathode rays were just like light rays. But there is one important difference – cathode rays can be bent by a magnet, light rays cannot.

Figure 5 shows what happens when a magnet is brought close to a cathode ray tube. The beam of electrons *bends*. The magnetic field pushes or *deflects* the electrons off course.

Using an electric field   Electrons carry a negative charge. They are attracted by a positive charge, and pushed away or repelled by a negative charge. Figure 6 shows what happens when two metal plates inside a cathode ray tube are *charged*. The beam of electrons is pulled towards the positive plate and pushed away from the negative plate. The beam bends upwards. If the positive plate were underneath, the beam would bend downwards.

Moving sideways   The beam of electrons in a cathode ray tube can also be bent sideways. The two charged metal plates need to be on each side of the tube. The beam bends towards the positive plate and is deflected off course. With the + plate on the left the beam moves towards the left. With the + plate on the right the beam moves off to the right.

These charged metal plates can be used to move cathode rays up or down, and from side to side. They are called *deflecting plates*.

Cathode-ray oscilloscopes

In the next section you will see how cathode ray tubes are used in black and white television. But first, a look at another instrument, one of the most useful scientific instruments ever made. It is called a *cathode-ray oscilloscope*, a C.R.O. for short. Figure 7 is a simple drawing of the main parts inside a C.R.O.

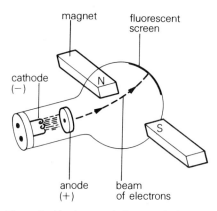

Figure 5   The beam of electrons is bent off course by a magnetic field.

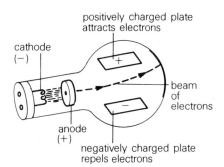

Figure 6   The beam of electrons is bent off course by an electric field.

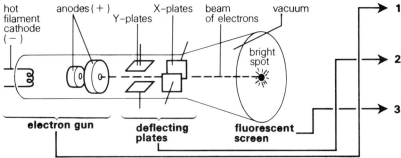

Figure 7   The main parts inside a C.R.O. are the electron gun, the deflecting plates, and the fluorescent screen.

1. The **electron gun** 'fires' a stream of fast-moving electrons. It consists of a hot cathode and an anode to attract the electrons.

2. The **deflecting plates:** the Y-plates can move the beam up and down, the X-plates can deflect it from side to side.

3. The **fluorescent screen** makes a bright spot where the electrons hit it.

Using a C.R.O. Cathode-ray oscilloscopes have very many uses. Every physics laboratory should have one. One of its uses is for displaying the pattern of a musical note on the screen.

A microphone is connected to the oscilloscope. When a sound wave reaches it, the microphone sends tiny electric currents to the C.R.O. This sets up a tiny voltage across the deflecting plates inside. As the voltage changes, the beam moves up and down to display the sound wave. Figure 8 shows the patterns on the screen with different notes. Only a tuning fork makes a perfect 'wave pattern' on the screen. A piano and a violin make a slightly rougher pattern.

Black and white television

A black and white television is a bit like a C.R.O. It has a fluorescent screen at the front (that you watch) and a cathode ray tube at the back. The beam of electrons makes a spot of light on the screen. This spot 'sweeps' quickly backwards and forwards across the screen to make *lines*. At the same time the spot is pulled *down* the screen. So the television picture is made up of hundreds of lines, 625 in all, as Figure 9 shows. The picture changes as it gets signals from the TV aerial. These signals control the brightness of the spot on different parts of the screen. Some spots will be black, others brilliant white, others inbetween. When you watch black and white television you are seeing thousands of dots of light in lines that are sweeping across the screen. These bright and dark dots make up the picture. You see 25 different pictures every second.

The picture on a colour TV is made up of thousands of red, green, blue, and white strips. These tiny strips build up a colour picture.

Using television Television is not just used for entertaining people. Other important uses are looking for traffic jams and spotting thieves.

low pitch
(low frequency)

high pitch
(high frequency)

a quiet note

same pitch
but louder

violin

piano

Figure 8 Different notes make different patterns on a C.R.O.

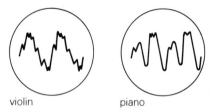

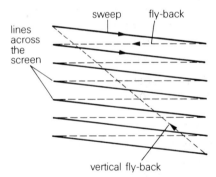

Figure 9 A black-and-white television picture is made up of 625 lines on the screen.

Summary

1 Extremely hot wires, or filaments, can give off free electrons. These free electrons can be made to travel through a vacuum in a cathode ray tube.

2 Cathode rays can be deflected by a magnet or by an electric field.

3 Cathode ray tubes are used in the cathode-ray oscilloscope (C.R.O.) and in televisions.

Exercises

1 Copy out and fill in the blanks:
Electrons can travel through a _____ in a _____ ray tube. The main parts of a _____ ray tube are the fluorescent _____ , the hot filament or _____ , the positive _____ , and the deflecting _____ . Beams of electrons are often called _____ rays. They can be deflected by a _____ or by an electric _____ .

2 What force keeps an electron in orbit around the nucleus? Describe how an electron can become 'free'.

3 Draw a simple diagram of a cathode ray tube and label: the cathode, the vacuum, the anode, and the screen.
What is special about the screen?

4 What are cathode rays?

5 What are the two ways of bending cathode rays?

6 What is a black and white television picture actually made up of? How is it different from a colour picture?

Whirlpools in space

Gateways to another universe, collapsed stars, invisible 'jaws' in outer space, whirlpools from which nothing can escape – black holes might be all of these things. The gravitational field of a black hole would be so strong that it would make the Sun's gravity seem weak. Black holes could swallow up the Universe, help man to reach the stars, or provide Earthlings with all the energy they will ever need. Which will happen first?

The story of black holes

A Frenchman called Pierre Laplace first suggested that outer space could be full of black holes. That was in 1795! He knew that every object in our Universe has a gravitational field.

Pierre Laplace first suggested, in 1795, that outer space could be full of black holes.

It is gravity that keeps people on Earth. But people can escape from the Earth's gravity. To escape completely a rocket has to reach a speed of more than 40 000 km/hour. This is called the Earth's *escape velocity*.

The Earth and its gravitational field are tiny compared with some of the stars in the Universe. Some massive stars have such strong gravity it would be impossible to escape from them. Their escape velocity would be higher than the speed of light, said Laplace. Light itself could not escape. The star would be *invisible*. The space around the star would be a 'black hole'.

Einstein About 120 years later a young man called Albert Einstein came up with some new theories. One of Einstein's important ideas was this: *nothing can travel faster than light*. So nothing can escape from a black hole. Escape is impossible.

Later in his life Einstein showed that black holes are *possible* – provided that matter can be squeezed into a small enough space . . .

How black holes are made

The Earth's mass is 6 million million million million kilograms. It is 12 700 km across. If all the matter in the Earth were squeezed until it was only 2 cm across it would become a tiny black hole. What about the Sun?

   The Sun would need to shrink from being 1 400 000 km across, to 6 km across! No light could escape from it then. The Sun would be a black hole.

Astronomers think that a *planet* could never become a black hole. But some *stars* could. Strange things can happen to a star . . .

The life of a star

Stars 'live' for millions of years. At the end of their lives they blow up in an enormous explosion. Lighter, less massive stars explode outwards. But heavier stars, three times as heavy as the Sun, explode inwards. They *implode*.

These stars collapse under their own enormous weight. They become smaller and smaller. They take up less and less space. Their mass is squeezed into a smaller and smaller volume. They become *denser* and *denser* as the collapsing star just goes on collapsing.

Neutron stars Some stars are squeezed into a *neutron star*. The atoms inside it destroy one another. Positive protons (+) are squeezed onto negative electrons (−). The + and − cancel each other out. Then the matter inside the star is *neutralised*. Only neutral particles, neutrons, are left. These are squeezed together to make the densest star of all, the neutron star. It may be only 20 km across. A sugar-cube sized piece of matter from it would have a mass of 100 million tonnes!

The Crab Nebula is the remnant of a star's explosion in 1054. At its centre is a neutron star which is all that is left of the original giant star that exploded.

But a neutron star can collapse even more. The star could keep on collapsing, getting smaller and smaller, until it occupied almost no space at all. It would finally become a black hole. All the matter of the original star would be contained in a tiny point in space. The region in space all round this point would have enormous gravity. Nothing could escape.

Do black holes exist?

Astronomers believe that some stars have already become black holes. The trouble is they cannot be seen! They are invisible because light cannot escape from them. But they can be *felt*. The pull of gravity close to a black hole may be as strong as 5000 million Suns.

Many stars in space are found in pairs. Each star in the pair spins around the other one. But some stars in space have been seen which go round and round, yet they seem to be on their own. Astronomers believe that perhaps they are being kept in orbit by the gravity from a black hole. This black hole is invisible. Yet the star is in orbit around it, just as the Earth orbits the Sun. If the star stopped moving it would be sucked into the black hole.

Using black holes

Endless energy? Black holes are dangerous. But one day they might be useful as an endless source of energy . . .

Moving objects near a black hole can be kept orbiting round and round, like a giant whirlpool. The black hole itself spins at a terrific speed. Scientists have already dreamed up projects for using this moving energy. A single black hole could provide 1000 Earth-like civilisations with enough energy to last until the end of the Universe.

But there are at least two problems: reaching a black hole, since the nearest one is at least ten million million kilometres away; and using its energy without being swallowed up by it.

It seems impossible now. But ten years before the first atomic bomb a famous physicist said that no atom bomb could ever be made.

Gateways to another galaxy?

Some astronomers believe that one day people will actually *build their own* black holes, which could be used to travel to other parts of the Universe. A brave astronaut would allow his or her spacecraft to be sucked into the whirlpool of a black hole. He would vanish from sight. But some scientists believe that the spacecraft could appear again – in a completely different part of the Universe. It would re-emerge through a *white hole*, they say.

No-one has yet seen a white hole. But some exploding galaxies, called Seyfert galaxies, are thought to be giant white holes which perhaps pour matter back *into* the Universe as quickly as black holes are taking it *out*.

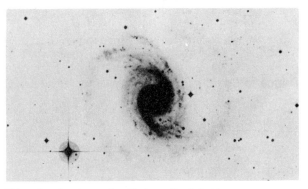

This galaxy has been classified as a Seyfert galaxy.

Space travel using black and white holes seems as unlikely nowadays as reaching the moon must have seemed to Laplace in 1795. But a few astronomers believe that black holes are mankind's only chance of reaching the stars or other galaxies.

Topic 6 Exercises

More questions on fields

1 This diagram shows how a signal can be made to move up when a current goes through a coil. Explain how it works. Why is iron used instead of steel?

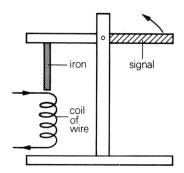

2 Explain what is meant by 'escape velocity'. Why is the escape velocity on Earth larger than it is on the moon?

3 Find out some of the uses of a cathode-ray oscilloscope (C.R.O. for short) in medicine, science, and industry.

4 Imagine you are in a completely dark room. You have three identical bars except that one is made of copper, the other of iron, the third of magnetised steel. How could you find out which is which?

5 Is it possible to magnetise a piece of steel so that it has no magnetic poles?

6 Find out why magnets are always stored with metal 'keepers' attached to their ends.

7 Copper is not attracted by a magnet. But when a copper pendulum swings between the poles of a magnet it slows down very quickly. Explain why.

Things to do

1 Defying gravity

You need: a bucket of water, some old clothes, and a hat.
Go outside and swing the bucket over your head in a circle. If you swing it quickly enough the water stays in the bucket. (If you don't, you get wet.) The water appears to be 'weightless', just like an astronaut in orbit around the Earth.

2 An 'orbiting' coin

Try placing a coin on the turntable of an old record player. At slow speeds the coin goes round in 'orbit' without flying off. The coin does not fly off because of friction (just as planets are kept in their orbits by the Sun's gravity). Speed up the turntable. The coin will fly off in a straight line.

3 Fishing with a magnet

You need: a bowl of water, cardboard, scissors, paper clips, string, and a magnet.
Cut out some cardboard fish. Attach just the right number of paper clips to each fish so that it stays suspended in the water – it neither floats nor sinks. Tie the magnet onto the string. Dangle the magnet in the water until you catch a fish. (This will amuse your little brother or sister for hours.)

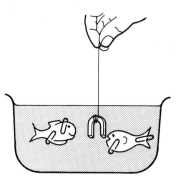

4 If pigs could fly . . .

You need: a paper clip, a magnet, some thread, an empty shoe box, paper, scissors, and sticky tape.
Tie the thread to the paper clip. Draw a figure (e.g. a pig or a cat) on the paper, then cut it out and tape it to the paper clip. Now fix the end of the thread to the bottom of the shoe box, so that the figure doesn't *quite* reach the top. Hide the magnet behind the lid of the box, as shown, and hold the figure just under it. Let go. To someone watching, the figure appears to float in mid air. Move the magnet around slightly—the figure seems to defy gravity.

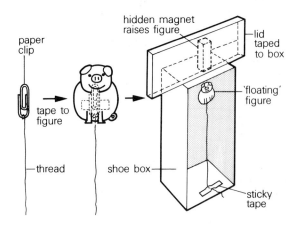

Index

Acknowledgements

Associated Press, p.3 (bottom);
Australian Information Service, London, p.84;
Barnaby's Picture Library, p.3 (centre), p.19 (top), p.33 (top and centre),
    p.37 (bottom), p.79, p.95, p.100 (bottom), p.110 (top), p.130 (bottom),
    p.139, p.147;
BBC Hulton Picture Library, p.104, p.121;
Biophoto Associates, p.3 (top), p.17 (top), p.83, p.91 (bottom), p.98,
    p.137 (right), p.140;
Dr Tony Brain/Science Photo Library, p.14 (bottom);
British Airways, p.62;
British Leyland, p.143 (bottom);
British Nuclear Fuels Ltd., p.131, p.132 (left), p.133 (right and top left);
Camera Press, p.2 (centre), p.31, p.37 (top), p.43 (bottom), p.91 (top);
Camera Press (Tass) London, p.132 (right).
Central Electricity Generating Board, p.35;
Central Office of Information, p.7, p.58 (bottom), p.119 (top);
Eastern Counties Newspapers, p.9 (left);
ESA/Science Photo Library, p.2 (bottom);
Gemeentemuseum, Escher 'Waterfall', Collection Haags, c. SPADEM, p.9
    (right);
Griffin and George, p.86, p.137 (bottom);
AERE Harwell, p.48 (right);
Hawker Siddeley/Science Photo Library, p.103 (bottom);
Italian State Tourist Office, p.80;
Keymed Industrial, p.113 (right);
Keystone Press Agency, p.57, p.119 (bottom), p.146 (top);
Loncraine Broxton & Partners Ltd., p.55;
Marley Waterproofing Products, p.13 (bottom);
Mary Evans Picture Library, p.1 (Galileo, Brown, Copernicus), p.8, p.17
    (bottom), p.146 (bottom), p.150;
Milton Keynes Development Corporation (John Walker), p.39;
NASA/Science Photo Library, p.28 (both), p.135;
National Center for Atmospheric Research/Science Photo Library,
    p.21 (top);
OUP archives, p.14 (top), p.100 (top);
Penrose, L. S. and R., ('Impossible Objects: A Special Type of Illusion',
    British Journal of Psychology, No. 49, 1958, p.31), p.9 (drawings in
    right column);
Philips Electronics, p.87;
Popperfoto, p.1 (Einstein), p.11, p.18, p.22 (bottom), p.58 (top), p.89, p.107
    (top), p.110 (bottom), p.112;
Rex Features Ltd., p.144;
Royal Astronomical Society, p.90, p.103 (top);
Royal Astronomical Society (Hale Observatories), p.29 (top right);
Royal Observatory, Edinburgh, p.29 (bottom right), p.151 (right);
Science Museum, p.1 (Dalton and Newton), p.25, p.27 (centre), p.81,
    p.116, p.148;
Science Photo Library (Hale Observatories), p.151 (left);
Shell Photographic Service, p.48 (left);
Smiths Industries, p.75;
Swiss National Tourist Office, p.2 (top), p.19 (bottom), p.23, p.33 (bottom),
    p.115, p.127;
J. Tabberner, p.22;
J. Thomas, p.24, p.27 (top), p.130 (centre);
Topham Picture Library, p.13 (top), p.49, p.110 (centre), p.113 (left),
    p.143 (top);
United Kingdom Atomic Energy Authority, p.133 (bottom left);
University of London, Courtauld Institute of Galleries, p.107 (bottom);
University of Manchester, p.1 (Rutherford);
US Air Force (Aerospace A.V. Service), p.29 (left);
R. Winstanley, p.12 (centre), p.21 (centre), p.32 (top and bottom), p.43
    (top), p.56, p.59, p.92, p.128, p.130 (top), p.141, p.143 (bottom).